EUREKA!

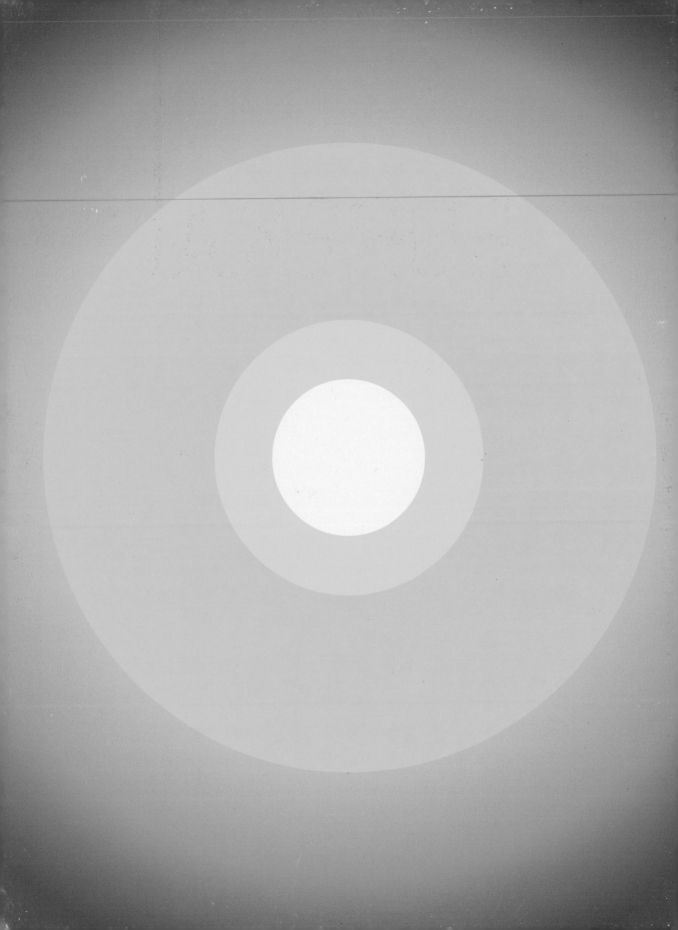

EUREKA!

An Infographic Guide to Science

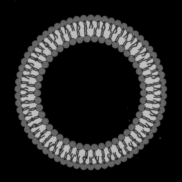

Tom Cabot

WILLIAM
COLLINS

For Megan, Edie and Rose

William Collins
An imprint of HarperCollinsPublishers
1 London Bridge Street
London SE1 9GF

WilliamCollinsBooks.com

First published by William Collins in 2016

© Tom Cabot, 2016

A catalogue record for this book is available
from the British Library.

Set in ITC Conduit, designed, illustrated and produced by
Tom Cabot/ketchup

Printed in Hong Kong by Printing Express

ISBN 978-0-00-812936-1

All reasonable efforts have been made by the author to trace the copyright
owners of the material quoted in this book and of any images reproduced
in this book. In the event that the author or publishers are notified of any
mistakes or omissions by copyright owners after publication of this book,
the author and the publishers will endeavour to rectify the position
accordingly for any subsequent printing.

CONTENTS

VISUAL CONTENTS

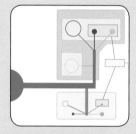
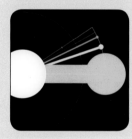

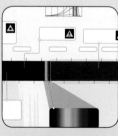

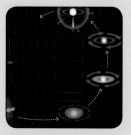

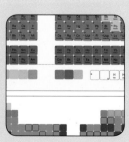
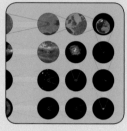
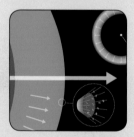

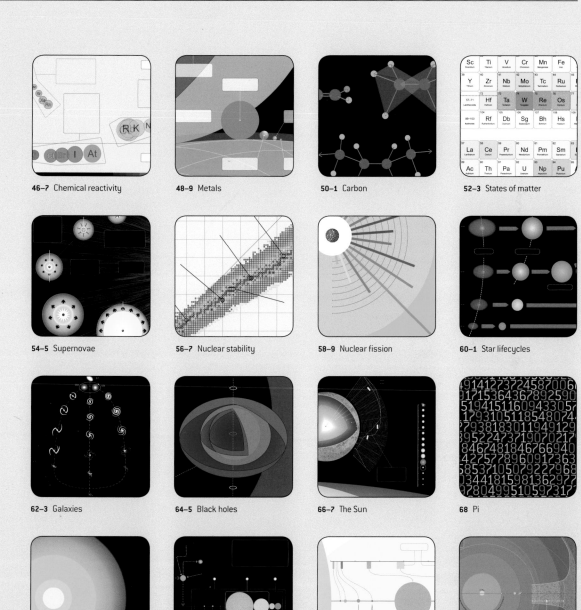

80–1 The Moon

82–3 The Earth

84–5 The lithosphere

86–7 Plate tectonics

88–9 Earthquakes

90–1 The atmosphere

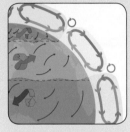

92–3 Atmospheric circulation

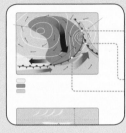

94–5 Meteorology

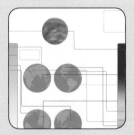

96–7 Climate extremes

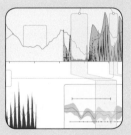

98–9 Solar activity and climate

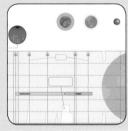

100–1 Snowball Earth

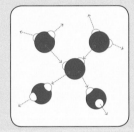

102–3 Water

104–5 Organic chemistry

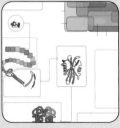

106–7 Proteins

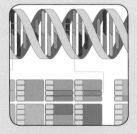

108–9 DNA coding

110 Earth

112 LIFE

116–17 Abiogenesis

118–19 First Life

120–1 Protocell

122–3 Cell membranes

124–5 Kingdoms of life

126–7 Eukaryotes

128–9 Mitochondria

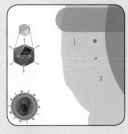

130–1 Viruses

132–3 The Great Oxygenation

134–5 Photosynthesis

136–7 Metabolic pathways

138–9 Enzymes

140–1 The immune system

142–3 Chromosomes

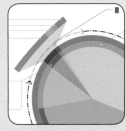

144–5 Cell reproduction

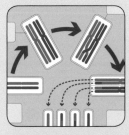

146–7 Sexual reproduction

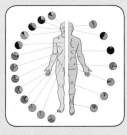

148–9 Bacterial microbiome

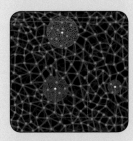

150–1 Single-cell life

152–3 Precambrian life

154–5 Bilaterians

156–7 Mass extinctions

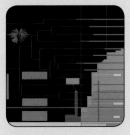

158–9 The Cambrian explosion

160–1 The chordates

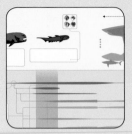

162–3 Fish

164–5 Marine habitats

166–7 Gills and lungs

168–9 Invasion of the land

170–1 The Carboniferous

172–3 Arthropods

174–5 Amniotes

176–7 Dinosaurs

178–9 Trees and forests

180–1 Pterosaurs

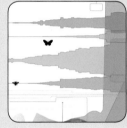

182–3 Co-evolution

184–5 Vertebrate vision

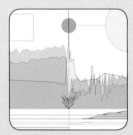

186–7 Chicxulub

188–9 Bird migration

190–1 The rise of the mammals

192–3 Gestation

194–5 Evolutionary return

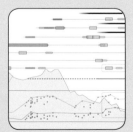

196–7 Grasslands

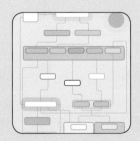

198–9 Biosphere and carbon

200 Extinct mammalian RNA

202 HUMAN

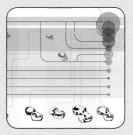

206–7 The primates

208–9 Early humans

210–11 The Human diaspora

212–13 Human anatomy

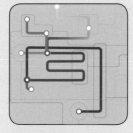

214–15 Digestion

216–17 Circulation

218–9 Hormones

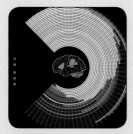

220–1 Language

222–3 Brain evolution

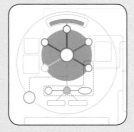

224–5 Perception

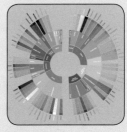

226–7 Taste

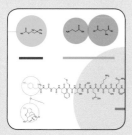

228–9 Neurotransmission

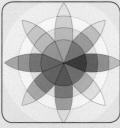

230–1 Emotion

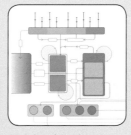

232–3 Learning and memory

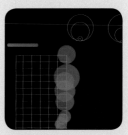

234–5 Computing

236–7 Artificial intelligence

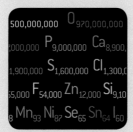

238–9 Human molecule

From the Big Bang to Artificial Intelligence

The history of science is a human story, for science is a human endeavour. Science is the systematic study of the structure and behaviour of the physical and natural world through observation and experiment. The vast majority of the objects of traditional scientific inquiry existed before humankind. Looking from the point of view of the Universe, if it were a sentient entity, the history of science is almost exclusively **not** a human story. We might well seem like the full point at the end of an extended footnote in an appendix to the final chapter of the story. If the 13.8 billion years since the birth of the Universe were compressed into a year, the hundred thousand or so years since humans walked the Earth would account for the last three minutes and fifty seconds before the chimes bringing in the New Year.

The alternative history of science is the history of the Universe: a history that takes as its starting point the most fundamental building blocks of all materials and follows how their self-assembly into ever more complex and fertile patterns, with extraordinary emergent qualities, describes and takes in every aspect of the material Universe currently studied, wondered about and understood.

At the start the Universe was pure **maths** then pure energy and force ... **physical** matter emerged then elemental matter ... hydrogen and helium. Free hydrogen, a proton orbited by an electron, expressing its new **chemical** properties, joined with another hydrogen atom and became the first molecule. Gravity, by a staggering distance the weakest force in nature, slowly assembled these molecules into vast stellar nuclear reactors where immense heat and pressure began to cook up the full gamut of elemental possibilities, aided by those stars' fiery and explosive ends.

Stars gave birth to new stars and planets were born, new physical **planetary** processes and new chemical compounds emerged: dihydrogen monoxide, otherwise known by its common name of 'water', became prevalent. And it was the proximity of water and other physical and chemical environments that led to perhaps the greatest material conjuring trick of relevance to us, the emergence of nature, **life** and natural science from pure chemistry. The story of the evolution of the human species from bacterial slime is both glorious and saddening in equal measure as we stand on the verge of wrecking our cradle. While the almost Platonic beauty of the physical Universe staggers with scale and majesty, the natural world amazes with intricacy, fiendish invention, quirk, and sheer dogged determination to express the unimaginable in the drive to thrive.

With humankind came the drive to describe, understand and symbolise the world around us, initially through art and ritual and then through methodology and measurement, the advent of language enabling the rapid and explosive evolution of thought. Just as life introduced replication, history, adaptation and development on a natural scale to the Universe of chemistry, so human linguistic intelligence introduced a more radical mastering of the natural environment on a new timescale. In the space of five thousand years, and perhaps more radically in the last two hundred, humans have not just found means to symbolically duplicate the entire knowable Universe and its history, but to use the resultant model to master and potentially irrevocably change the entire terrestrial system. For the first time in the history of the solar system, the future of a planet is in the hands of a single living species.

This book charts that process through the disciplines – often pioneered by scientists as wayward as any of our great artists – of physics and chemistry, DNA and biology, the brain and technology ... and on to the merging of human technology with human intelligence – the journey from pure energy to pure intellect.

Powers of ten: from Trafalgar Square to the largest structures in the observable Universe

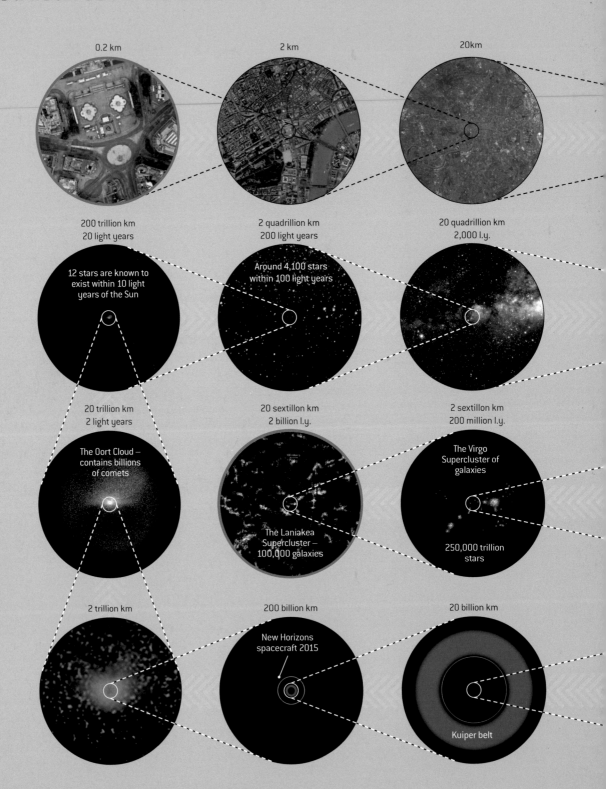

0.2 km

2 km

20km

200 trillion km
20 light years

12 stars are known to exist within 10 light years of the Sun

2 quadrillion km
200 light years

Around 4,100 stars within 100 light years

20 quadrillion km
2,000 l.y.

20 trillion km
2 light years

The Oort Cloud — contains billions of comets

20 sextillon km
2 billion l.y.

The Laniakea Supercluster — 100,000 galaxies

2 sextillon km
200 million l.y.

The Virgo Supercluster of galaxies

250,000 trillion stars

2 trillion km

200 billion km

New Horizons spacecraft 2015

20 billion km

Kuiper belt

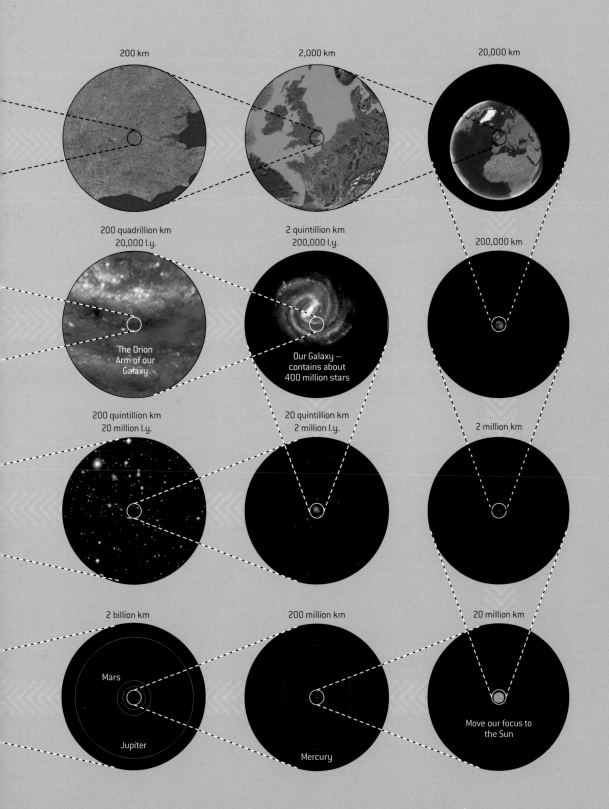

200 km

2,000 km

20,000 km

200 quadrillion km
20,000 l.y.

The Orion
Arm of our
Galaxy

2 quintillion km
200,000 l.y.

Our Galaxy —
contains about
400 million stars

200,000 km

200 quintillion km
20 million l.y.

20 quintillion km
2 million l.y.

2 million km

2 billion km

Mars

Jupiter

200 million km

Mercury

20 million km

Move our focus to
the Sun

UNIVERSE

Universe

In the twentieth century our sense of the Universe changed as radically as it did during the Renaissance, when humans finally threw away the conviction that the Earth lay at the centre of everything. We discovered that stars and galaxies aren't just moving away from each other like debris after an explosion; space itself is expanding. Even the space between your pupils and the words on this page is expanding. And, as our measurement and observation of space refined, we discovered that this universal inflation is accelerating, that the Universe was once smaller and hotter – very small and very hot. At a certain point, as the clock is turned back as far as it will go, the known laws of physics cease to be valid and the entire Universe tends to a singularity at a finite point in time – a theoretical point, however, of infinities. Infinitely massive, infinitely small, infinitely hot. A time when all the known laws of physics break down. The birth of everything we know.

As of 2015, our most precise measurements, the latest and best from the European Space Agency's Planck space observatory, date this moment to 13,799,000,000 years ago ... plus or minus 21,000,000 years. What followed that moment was a mind-bending expansion, of a scale and speed that is hard to conceptualise. In 0.00000000000000000000000000000000015 of a second the Universe expanded from being smaller than a proton to something the size of a grapefruit – an increase of the order of 10^{26} times (1 followed by 26 zeros). From there the expansion and

After about a millionth of a second the Universe was perhaps the size of our solar system and had cooled to around ten trillion degrees kelvin. At this temperature quarks coalesced to form protons and neutrons: matter, as we recognise it, the stuff of tables and butterflies, stars and flowers, books and sculpture, began to form. After a second the Universe was probably around ten thousand times larger than our solar system … and so the expansion and cooling continued.

For nearly 400,000 years the Universe glowed like the plasma of a candle flame, but then, as the temperature dropped below three thousand degrees kelvin, almost in a moment space became transparent — light was free to speed through space and time. Matter swirled and coalesced. Space, left dappled by minuscule effects at its birth, began to mottle and clump. Vast clouds of hydrogen and helium started to fall in on themselves, twisting as they did so. Delicate, but persistent, gravity eventually created huge swirling discs of gas, heating and compressing under pressure of mass. Thus the first stars were born. One by one these thermonuclear lights switched on through the firmament: small furnaces burning hydrogen and helium and creating the first heavier elements … carbon, oxygen, silicon, iron.

The biggest stars burn brightest and die the quickest — collapsing as they expire and exploding in massive supernovae. The violence and heat of these explosions forged new, heavier elements and scattered them to the heavens, seeding interstellar space with huge molecular clouds containing a variety of elements. These new star nurseries collapsed inwards again, spinning to form stars surrounded by discs rich in dusts and ices: metals, organic compounds, water!

And so the cycles of star and planet formation continued. Starlight, radiation energising elements to react and form ever-more complex substances. Rocky planets awash with water, icy gas giants with seas of methane. The nuclear Universe overlaid by the Universe of chemistry and geophysics: the nature of atoms determining how they combine and recombine. The unique and almost magical properties of water allowing the evolution of ever-more complex chemistry …

The geometry of the Universe

It's hard to visualise the 'shape of space' as we are inside it. We are constrained by perceptual systems that operate solely within that space. We can't pull our head out of space and regard its dimensions. But there is no reason why space should be 'flat', and the shape of space has fundamental consequences for the future of the Universe: will the Universe continue to expand forever, or will it slow its expansion and eventually collapse back on to itself in a 'Big Crunch'?

To think about the geometry of three dimensions, let us scale down to two. Imagine a cube of empty space – it could be a metre across, or 90 billion light years across.

If space is flat, if you set out from a point (Z) and head in a straight line a set distance before changing direction, it would take you three 90° turns to come back to where you started. Obviously, in three dimensions you could turn up and down as well as left and right and you might have to make more turns to return to Z, but the principle pertains. A particle (say a photon of visible light) travelling in a straight line without changing direction, will go from A to B and onwards towards infinity. Two particles travelling in parallel will never meet.

But if you take that flat space and curve it so the three dimensions are closed, with no edges, just like the two-dimensional surface of a sphere, different things happen when you move through that space. Starting at Z you might only need to turn two corners to return to where you started. Parallel paths can converge. A photon of light travelling in a straight line from point A might eventually return to its point of origin. Light from our Sun could return from deep space to be perceived as a distant star.

The Universe seemingly began as a singularity – a point with no dimensions. Everything that we know to exist, all space and time, emerged in an unimaginably vast and rapid expansion. Space itself was born. That expansion continues to this day with stars and galaxies racing apart as the fabric of the Universe grows – even the space between your eyes and this page.

The emergence of theories of relativity, where time was no longer regarded as a constant, required the conception of spacetime. Spacetime is a four-dimensional Euclidean space with the three (familiar) spatial dimensions plus time.

Just as two-dimensional surfaces have a shape – some flat, some convex, some concave – three-dimensional space can have varying geometry. Since the early twentieth century scientists have questioned the shape of space. Recent measurements of the Cosmic Background Microwave radiation have established, within an accuracy of 0.4 per cent that the curvature of space is flat.

Everywhere in the Universe was once somewhere ... or, rather, nowhere. The space you occupy was once at the centre of the Universe, as was every other point in the 91 billion light-year wide volume we now call home.

2-D 'net' of a 3-D cube:
a 3-D cube can be 'unfolded' to six 2-D squares

3-D 'net' of a 4-D cube (tesseract):
a 4-D cube can be 'unfolded' to eight 3-D cubes

Just as a 3-D cube can cast a 2-D shadow on a wall, in theory, a tesseract could throw 3-D shadows of itself into 3-D space.

It might well be that to fully understand the structure and strangeness of the Universe (Why is gravity so weak? Why is 95% of the visible Universe apparently missing?), we will have to acknowledge the existence of additional spatial dimensions that we can't fully perceive. Is the matter that makes everything we see a three-dimensional manifestation of the excitation of four-dimensional fields? Certainly, allowing for extra dimensions solves some of the remaining problems with the maths underlying our understanding of the quantum Universe.

3-D projection of a tessaract, or 4-D hypercube

The unfurling fundamental forces

The Universe: the first millionth of a second

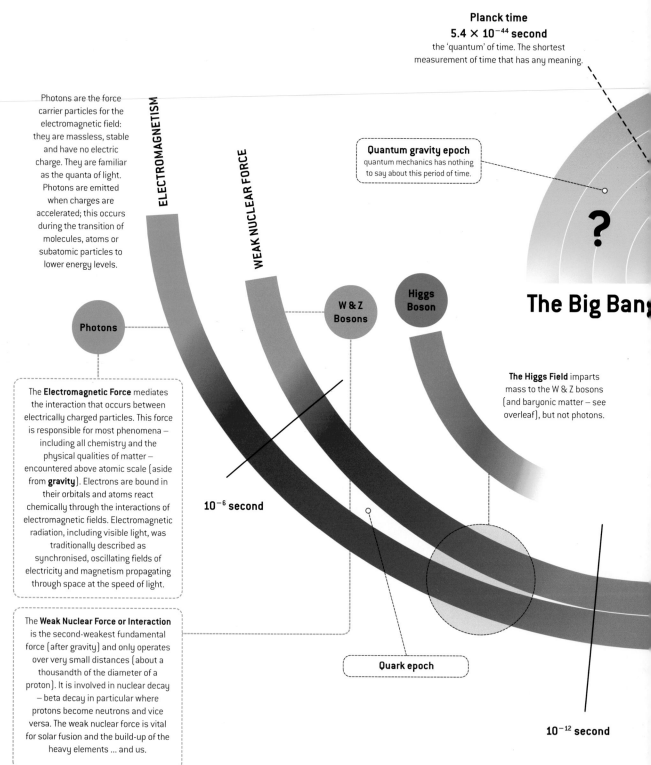

Planck time
5.4×10^{-44} second
the 'quantum' of time. The shortest
measurement of time that has any meaning.

Photons are the force
carrier particles for the
electromagnetic field:
they are massless, stable
and have no electric
charge. They are familiar
as the quanta of light.
Photons are emitted
when charges are
accelerated; this occurs
during the transition of
molecules, atoms or
subatomic particles to
lower energy levels.

ELECTROMAGNETISM

WEAK NUCLEAR FORCE

Quantum gravity epoch
quantum mechanics has nothing
to say about this period of time.

?

Photons

W & Z Bosons

Higgs Boson

The Big Bang

The **Electromagnetic Force** mediates
the interaction that occurs between
electrically charged particles. This force
is responsible for most phenomena –
including all chemistry and the
physical qualities of matter –
encountered above atomic scale (aside
from **gravity**). Electrons are bound in
their orbitals and atoms react
chemically through the interactions of
electromagnetic fields. Electromagnetic
radiation, including visible light, was
traditionally described as
synchronised, oscillating fields of
electricity and magnetism propagating
through space at the speed of light.

10^{-6} second

The Higgs Field imparts
mass to the W & Z bosons
(and baryonic matter – see
overleaf), but not photons.

Quark epoch

10^{-12} second

The **Weak Nuclear Force or Interaction**
is the second-weakest fundamental
force (after gravity) and only operates
over very small distances (about a
thousandth of the diameter of a
proton). It is involved in nuclear decay
– beta decay in particular where
protons become neutrons and vice
versa. The weak nuclear force is vital
for solar fusion and the build-up of the
heavy elements ... and us.

The graviton is a hypothetical particle that mediates gravity. If discovered, the graviton would unite quantum theory with relativity. Quantum gravity is largely a theoretical exercise as its effects are only thought to become apparent at close to Planck scale – a realm way beyond the capabilities of practically envisioned particle accelerators, like the Large Hadron Collider.

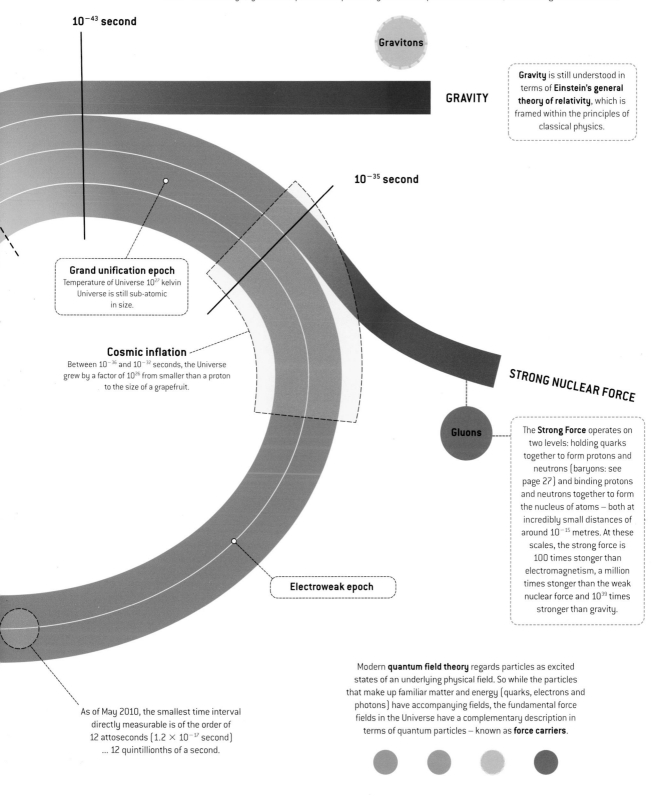

10^{-43} second

Gravitons

GRAVITY

Gravity is still understood in terms of **Einstein's general theory of relativity**, which is framed within the principles of classical physics.

10^{-35} second

Grand unification epoch
Temperature of Universe 10^{27} kelvin
Universe is still sub-atomic
in size.

Cosmic inflation
Between 10^{-36} and 10^{-32} seconds, the Universe grew by a factor of 10^{26} from smaller than a proton to the size of a grapefruit.

STRONG NUCLEAR FORCE

Gluons

The **Strong Force** operates on two levels: holding quarks together to form protons and neutrons (baryons: see page 27) and binding protons and neutrons together to form the nucleus of atoms – both at incredibly small distances of around 10^{-15} metres. At these scales, the strong force is 100 times stonger than electromagnetism, a million times stonger than the weak nuclear force and 10^{39} times stronger than gravity.

Electroweak epoch

As of May 2010, the smallest time interval directly measurable is of the order of 12 attoseconds (1.2×10^{-17} second) … 12 quintillionths of a second.

Modern **quantum field theory** regards particles as excited states of an underlying physical field. So while the particles that make up familiar matter and energy (quarks, electrons and photons) have accompanying fields, the fundamental force fields in the Universe have a complementary description in terms of quantum particles – known as **force carriers**.

Components of matter

The density of the Universe is made up of both energy and matter. Contrary to a popular belief following on from the most famous equation in modern physics, $E=mc^2$, energy and matter are not interchangeable expressions of the same thing.

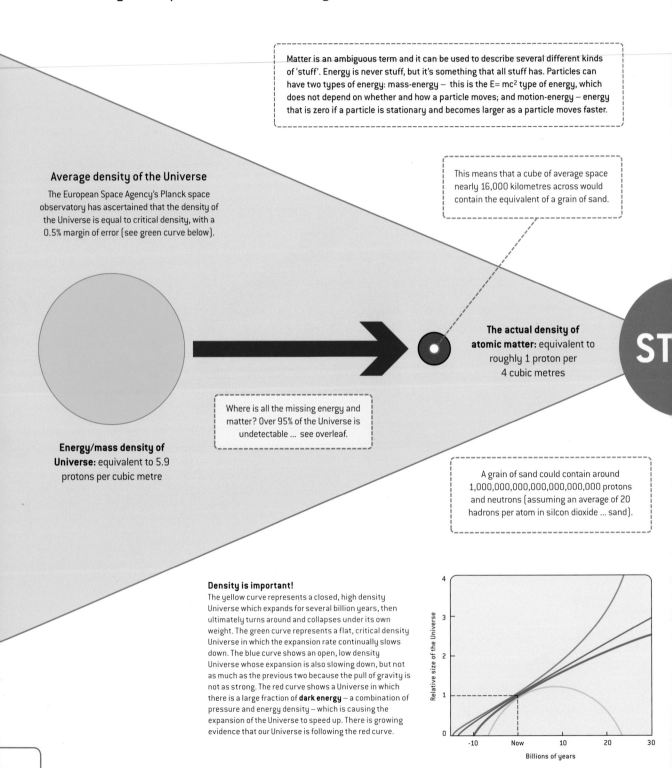

Matter is an ambiguous term and it can be used to describe several different kinds of 'stuff'. Energy is never stuff, but it's something that all stuff has. Particles can have two types of energy: mass-energy — this is the $E=mc^2$ type of energy, which does not depend on whether and how a particle moves; and motion-energy — energy that is zero if a particle is stationary and becomes larger as a particle moves faster.

Average density of the Universe

The European Space Agency's Planck space observatory has ascertained that the density of the Universe is equal to critical density, with a 0.5% margin of error (see green curve below).

This means that a cube of average space nearly 16,000 kilometres across would contain the equivalent of a grain of sand.

The actual density of atomic matter: equivalent to roughly 1 proton per 4 cubic metres

ST

Energy/mass density of Universe: equivalent to 5.9 protons per cubic metre

Where is all the missing energy and matter? Over 95% of the Universe is undetectable ... see overleaf.

A grain of sand could contain around 1,000,000,000,000,000,000,000 protons and neutrons (assuming an average of 20 hadrons per atom in silcon dioxide ... sand).

Density is important!

The yellow curve represents a closed, high density Universe which expands for several billion years, then ultimately turns around and collapses under its own weight. The green curve represents a flat, critical density Universe in which the expansion rate continually slows down. The blue curve shows an open, low density Universe whose expansion is also slowing down, but not as much as the previous two because the pull of gravity is not as strong. The red curve shows a Universe in which there is a large fraction of **dark energy** — a combination of pressure and energy density — which is causing the expansion of the Universe to speed up. There is growing evidence that our Universe is following the red curve.

Relative size of the Universe

4

3

2

1

0

-10   Now   10   20   30

Billions of years

The **stuff** of the Universe is all made from fields — the basic ingredients of the Universe — and their particles. All particles are ripples in fields and have energy. **Matter** can refer to atoms, the basic building blocks of what we perceive as the material Universe around us — fire, mountains, eyelashes, rockpools, salt — and indeed the components out of which atoms are made, including electrons and the protons and neutrons that make up the atom's nucleus. Or it can refer to what are called the elementary matter particles of nature: electrons, muons, taus, the three types of neutrinos, the six types of quarks — all of the types of particles which are not the force-carrier particles: the photon, gluons, the W and Z particles … and, potentially, the graviton (!) outlined on the previous spread.

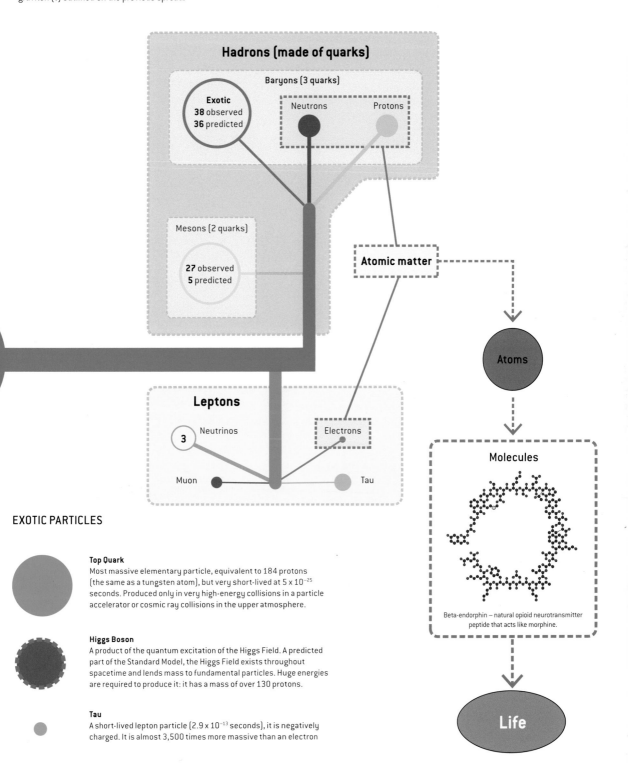

Hadrons (made of quarks)

Baryons (3 quarks)

Exotic
38 observed
36 predicted

Neutrons   Protons

Mesons (2 quarks)

27 observed
5 predicted

Atomic matter

Atoms

Leptons

3 Neutrinos

Electrons

Muon   Tau

Molecules

Beta-endorphin – natural opioid neurotransmitter peptide that acts like morphine.

Life

FF

EXOTIC PARTICLES

Top Quark
Most massive elementary particle, equivalent to 184 protons (the same as a tungsten atom), but very short-lived at 5×10^{-25} seconds. Produced only in very high-energy collisions in a particle accelerator or cosmic ray collisions in the upper atmosphere.

Higgs Boson
A product of the quantum excitation of the Higgs Field. A predicted part of the Standard Model, the Higgs Field exists throughout spacetime and lends mass to fundamental particles. Huge energies are required to produce it: it has a mass of over 130 protons.

Tau
A short-lived lepton particle (2.9×10^{-13} seconds), it is negatively charged. It is almost 3,500 times more massive than an electron

Matter and missing matter

It turns out that normal matter is made up of a dazzling array of
subatomic particles ... and then there's antimatter. But how do we
explain the fact that most of the matter and energy in the Universe
is invisible and, as yet, undetectable? Well, we can't. Yet.

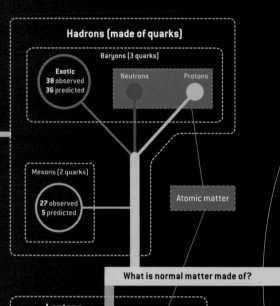

Dark matter – 26.8%

Normal matter – 4.9%

Hadrons (made of quarks)

Baryons (3 quarks)

Exotic
38 observed
36 predicted

Neutrons   Protons

Mesons (2 quarks)

27 observed
5 predicted

Atomic matter

What is normal matter made of?

Leptons

Neutrinos

3

Muon   Tau

Electrons

Electrons (along with neutrons and protons) are the other
main ingredient of normal matter. They are stable and
have a mass more than 1,800 times smaller than that of a
proton. There are three flavours of neutrino (they can
change in mid-flight) and they have a mass more than
500,000 times smaller than that of an electron. Muons
and tau particles are massive and very short-lived.

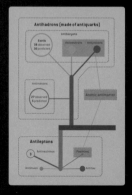

Antimatter

All subatomic particles have twin
anti-particles. Antimatter and
matter annihilate completely on
contact, releasing hugely
energetic photons.

The 94% of the mass of the visible Universe is made up of neutrons, protons and electrons. Nearly **ALL** science deals with this atomic matter. Exotic hadrons and mesons are all very unstable with half-lives generrlly between 10^{-10} and 10^{-24} seconds. Protons have a half-life greater than the age of the Universe ... free neutrons have a half-life of about 10 minutes, but, within the haven of the atomic nucleus, they are effectively stable.

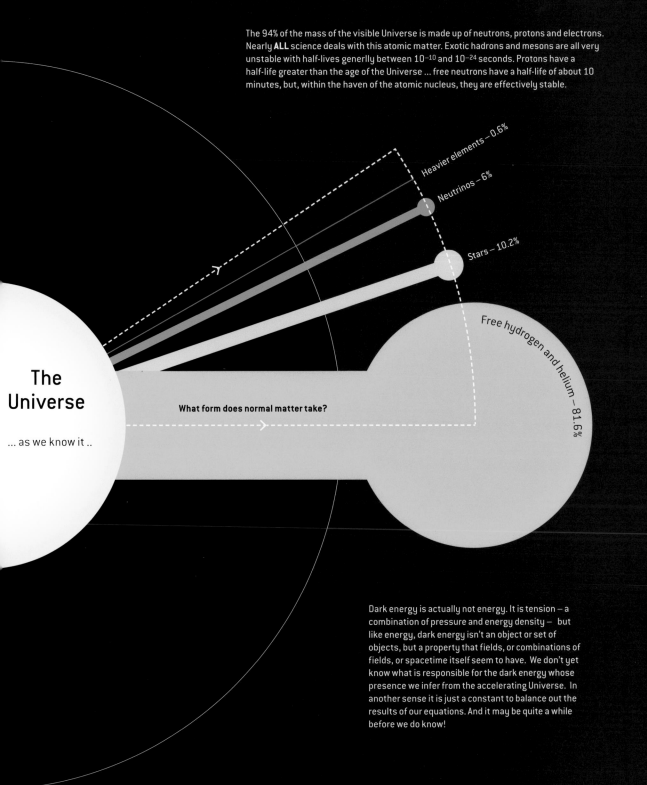

Heavier elements – 0.6%

Neutrinos – 6%

Stars – 10.2%

The Universe

... as we know it ..

What form does normal matter take?

Free hydrogen and helium – 81.6%

Dark energy is actually not energy. It is tension – a combination of pressure and energy density – but like energy, dark energy isn't an object or set of objects, but a property that fields, or combinations of fields, or spacetime itself seem to have. We don't yet know what is responsible for the dark energy whose presence we infer from the accelerating Universe. In another sense it is just a constant to balance out the results of our equations. And it may be quite a while before we do know!

Dark energy – 68.3%

Big Bang to 380,000 years later

The radiation afterglow of the Big Bang is still all around us. If you de-tune a television, a few percent of the snow-like static on screen is the result of this ancient light.

Photon Epoch
After the majority of leptons are annihilated the Universe is dominated by radiation – gamma ray photons that interact with the remaining charged protons, electrons and, eventually, nuclei for the next 380,000 years. Light cannot travel far before it is scattered or absorbed – the Universe is glowing, foggy, opaque.

Hadron Epoch
Universe is a ten trillion kelvin – the quark–gluon plasma that composes the Universe cools enough to allow quarks to form hadrons, including baryons such as protons and neutrons. Matter and antimatter are initially in equilibrium.

Lepton Epoch
After the majority of hadrons are annihilated the Universe is dominated by leptons: mostly electrons and antimatter positrons (and the more exotic leptons ... tau, muon and the neutrinos.

COSMIC NUCLEOSYNTHESIS

1 microsecond (10^{-6} seconds)

Quark Epoch
Universe is ten million billion kelvin – space is filled with an ocean of free quarks and gluons.

1 second

10 seconds

100 seconds

Diameter of Universe	0.001312 light years	1.3 light years

About a thousand times the size of our solar system

Temperature drops below ten billion kelvin, hadron matter and antimatter annihilates – a slight excess of matter, so one in a billion protons and neutrons survive!

Temperature drops below five billion kelvin, leptons and antileptons annihilate, leaving a small amount of lepton residue

Matter-dominated Universe (from 60,000 years)

Temperature of the Universe falls to one billion kelvin – atomic nuclei can begin to form. Protons (hydrogen ions) and neutrons begin to combine into atomic nuclei in the process of nuclear fusion. These consisted only of the nuclei of the simplest chemical elements: mostly **hydrogen** and **helium**. After only about seventeen minutes the temperature and density of the Universe has fallen to the point where nuclear fusion cannot continue. At this time, there is about three times more hydrogen than helium, by mass, and only trace quantities of other nuclei such as **lithium**.

Looking back at events from 'Now'

At the beginning, the Universe was zero size and has expanded away ever since. As we look out into space, the further we look the further back in time we look. The past 'moves away' from us, so the furthest back in time (now 29 billion light years away) is to the point where the Universe became transparent and light (photons) was able to travel through the intervening space to be detected by us. This earliest light is the source of what is called the **Cosmic Background Radiation**.

Infinity

Opaque Universe

Last scattering surface (380,000 years after Big Bang)

Transparent
Universe

Earth today

29 billion light years

Recombination

The temperature of the Universe drops below 3,000 kelvin – charged hydrogen nuclei can capture free electrons. Proportion of free electrons and protons to neutral hydrogen drops to a few parts to 10,000. Space becomes transparent and light is free to travel infinitely. The light from this time has been travelling through space ever since, and can be detected all around us. We can measure the afterglow of the Big Bang.

380,000 years after the Big Bang

90 million light years

Towards the present

Photon decoupling

The recombination of hydrogen and helium to form neutral atoms releases photons to travel freely through space. The temperature of 3,000 kelvin of this cosmic background radiation would have glowed a dull red. The source of this light is now so distant, and travelling away from us at such speed (because of the expanding Universe) that the light is redshifted all the way down to non-visible microwave radiation. This radiation – its nature, its texture – is now giving us insight into the earliest, infinitesimal moments and the very finest fabric of spacetime.

Mapping the Cosmic Microwave Background Radiation

Between 2001 and 2009, NASA's Wilkinson Microwave Anisotropy Probe mapped the cosmic microwave background radiation and, in the process, confirmed and extended the current Standard Model of cosmology. Subsequently, between 2009 and 2012, the European Space Agency's Planck Space Observatory continued and refined these observations. The mottled texture of minute temperature variations resulted from quantum fluctuations that occurred in the extraordinary inflation in the earliest moments after the Big Bang (10^{-32} seconds)! It is this texture that produced gravitational clumping of matter ... and thus stars and galaxies ... and us.

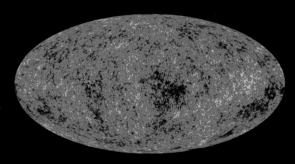

Electromagnetic radiation

All matter above absolute zero (−273° C or 0 kelvin) emits electromagnetic radiation – the wavelength of radiation is dependent on the temperature of the matter. The acceleration of charged particles (electrons and protons) produces electromagnetic radiation. These accelerations can be due to the thermal vibrations of atoms and atomic bonds within molecules, collisions between atoms, as well as the movement of electrons between energy states.

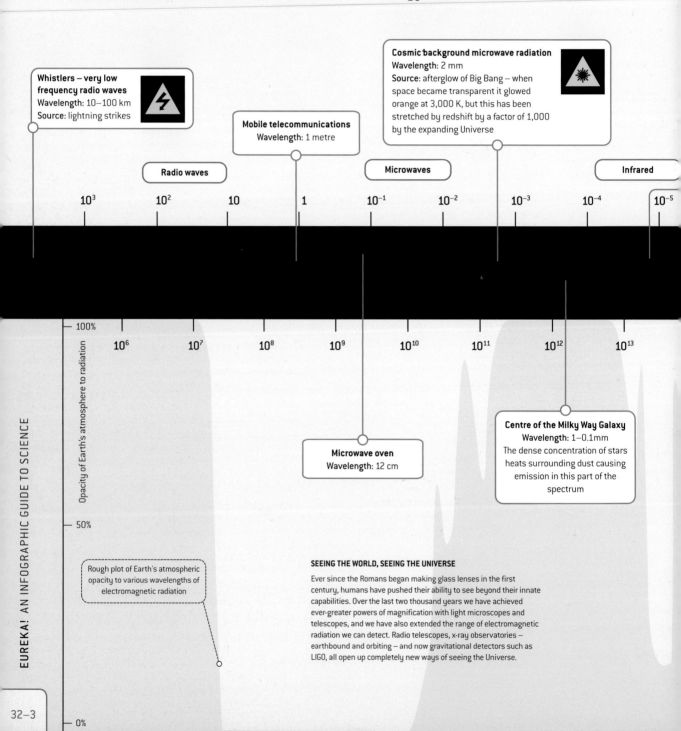

Whistlers – very low frequency radio waves
Wavelength: 10–100 km
Source: lightning strikes

Mobile telecommunications
Wavelength: 1 metre

Cosmic background microwave radiation
Wavelength: 2 mm
Source: afterglow of Big Bang – when space became transparent it glowed orange at 3,000 K, but this has been stretched by redshift by a factor of 1,000 by the expanding Universe

Radio waves

Microwaves

Infrared

| 10^3 | 10^2 | 10 | 1 | 10^{-1} | 10^{-2} | 10^{-3} | 10^{-4} | 10^{-5} |

Opacity of Earth's atmosphere to radiation

100%

| 10^6 | 10^7 | 10^8 | 10^9 | 10^{10} | 10^{11} | 10^{12} | 10^{13} |

Centre of the Milky Way Galaxy
Wavelength: 1–0.1mm
The dense concentration of stars heats surrounding dust causing emission in this part of the spectrum

Microwave oven
Wavelength: 12 cm

50%

Rough plot of Earth's atmospheric opacity to various wavelengths of electromagnetic radiation

SEEING THE WORLD, SEEING THE UNIVERSE

Ever since the Romans began making glass lenses in the first century, humans have pushed their ability to see beyond their innate capabilities. Over the last two thousand years we have achieved ever-greater powers of magnification with light microscopes and telescopes, and we have also extended the range of electromagnetic radiation we can detect. Radio telescopes, x-ray observatories – earthbound and orbiting – and now gravitational detectors such as LIGO, all open up completely new ways of seeing the Universe.

0%

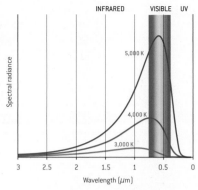

INFRARED VISIBLE UV

Spectral radiance

5,000 K

4,000 K

3,000 K

3 2.5 2 1.5 1 0.5 0

Wavelength (μm)

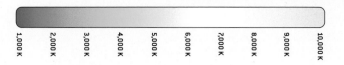

1,000 K 2,000 K 3,000 K 4,000 K 5,000 K 6,000 K 7,000 K 8,000 K 9,000 K 10,000 K

BLACK-BODY RADIATION AND INCANDESCENCE

Black-body radiation is the theoretically perfect thermal radiation emitted by a body. It is distinguished by a range of wavelengths with a distinct peak that correlates to its perceived colour. As an object is heated it will begin to glow at around 798 K (525 °C) – in the dark this is just faintly visible, appearing grey (because the human eye is sensitive only to black and white at very low intensities). As its temperature increases further its colour moves through dull red, orange, white – finally becoming blindingly brilliant blue-white, before passing into the realm of ultraviolet and x-rays.

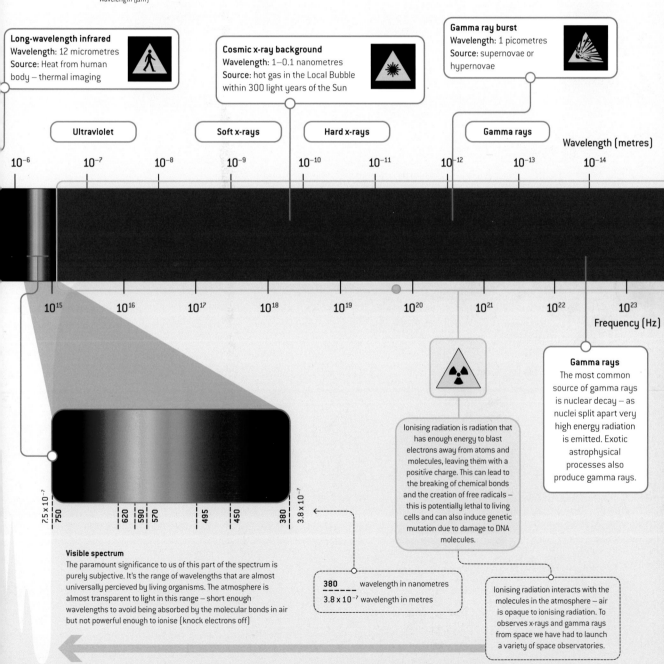

Long-wavelength infrared
Wavelength: 12 micrometres
Source: Heat from human body – thermal imaging

Cosmic x-ray background
Wavelength: 1–0.1 nanometres
Source: hot gas in the Local Bubble within 300 light years of the Sun

Gamma ray burst
Wavelength: 1 picometres
Source: supernovae or hypernovae

Ultraviolet

Soft x-rays

Hard x-rays

Gamma rays

Wavelength (metres)

10^{-6} 10^{-7} 10^{-8} 10^{-9} 10^{-10} 10^{-11} 10^{-12} 10^{-13} 10^{-14}

10^{15} 10^{16} 10^{17} 10^{18} 10^{19} 10^{20} 10^{21} 10^{22} 10^{23}

Frequency (Hz)

Gamma rays
The most common source of gamma rays is nuclear decay – as nuclei split apart very high energy radiation is emitted. Exotic astrophysical processes also produce gamma rays.

Ionising radiation is radiation that has enough energy to blast electrons away from atoms and molecules, leaving them with a positive charge. This can lead to the breaking of chemical bonds and the creation of free radicals – this is potentially lethal to living cells and can also induce genetic mutation due to damage to DNA molecules.

7.5×10^{-7} 750 620 590 570 495 450 380 3.8×10^{-7}

Visible spectrum
The paramount significance to us of this part of the spectrum is purely subjective. It's the range of wavelengths that are almost universally percieved by living organisms. The atmosphere is almost transparent to light in this range – short enough wavelengths to avoid being absorbed by the molecular bonds in air but not powerful enough to ionise (knock electrons off)

380 wavelength in nanometres
3.8×10^{-7} wavelength in metres

Ionising radiation interacts with the molecules in the atmosphere – air is opaque to ionising radiation. To observes x-rays and gamma rays from space we have had to launch a variety of space observatories.

Relativity

Classical Newtonian physics described gravity, mass and the movement of celestial bodies with amazing accuracy and was our best description of the cosmos for two hundred years. Force, mass and acceleration were all described with time and space as constants. Relativity transformed our view of the cosmos: now both space and time were variable; the speed of light (in a vacuum) became the limiting constant.

GRAVITY IN EINSTEINIAN PHYSICS

Relativity states that acceleration and the effects of gravity are indistinguishable. It is the local distortions of spacetime by massive objects, such as planets and stars, that deflect objects and cause them to fall into orbit. Gravity is not a force – it is a condition of the shape of space itself. An orbiting planet is travelling in a straight line relative to local spacetime, but as that space is curved, to an external observer the path of the planet is curved.

GRAVITATIONAL LENSING

One of the predictions of Einstein's theories of General Relativity was that large masses or concentrations of energy would distort the fabric of spacetime. The denser and more massive the object, the greater the distortion of spacetime. A very massive and localised galaxy cluster can create a deep well in the spacetime that intervenes between and observer on Earth and a very distant light source. Light from the distant object will be bent around the edges of that well – the apparent position of the distant object will move away from the centre of the nearer massive object.

Photons from distant object passing close to the nearer massive object are bent, creating a mirage image. Light passing on opposite sides will create separate mirage points of light.

Distant radiation source: bright galaxy or quasar many millions of light years away.

SPACETIME

Classical mechanics and physics operate within a fixed Euclidean three-dimensional space. Relativity uses a four-dimensional space and time 'manifold' (here represented as a two-dimensional plane). Around massive bodies time dilates, distorting spacetime – this distorts the path of moving bodies, be they objects with resting mass (a satellite) or photons (wave packets) of light with no resting mass.

Nearer massive object: galaxy cluster; multi-solar mass star; black hole – distorting space-time

Observatory on Earth

Gravity's spectrum

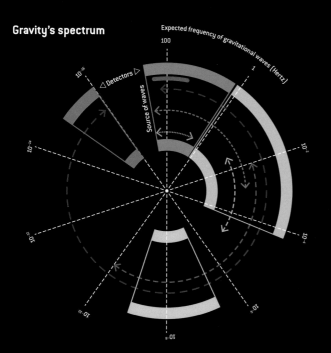

Expected frequency of gravitational waves (Hertz)

100

1

Detectors ▷

Source of waves

10⁻¹⁶

10⁻¹⁴

10⁻¹²

10⁻²

10⁻⁴

10⁻⁶

10⁻⁸

10⁻¹⁰

Potential source of gravitational waves

– – – – Inflation of the Universe, moments after the Big Bang (see page

– – – – Pairs of supermassive black holes orbiting around each other

- - - - - - Dense orbiting or inspiralling binary star or black-hole systems

— — — — Orbiting dense stars captured by supermassive black holes

.... Rotating neutron stars/violent hypernovae

Detectors built or planned

BICEP2 and other cosmic microwave background telescopes: gravity waves at birth of Universe polarises early light

Pulsar timing based on telescopic observation: gravity waves extract energy from binary star systems, causing inspiralling

Advanced LIGO, Virgo – earth-based laser interferometry: direct measurement of stretching and compression of spacetime

eLISA spacecraft – space-based, orbiting laser interferometry (possible launch 2032): direct measurement of strain in spacetime over millions of kilometres

GRAVITY WAVES

Einstein predicted that very large, violent mass events could create ripples in the fabric of spacetime – these would propagate through spacetime at the speed of light, carrying energy in the form of gravitational radiation. Just as accelerating charges propagate electromagnetic waves, accelerating masses produce gravity waves. A 50-year search culminated in the first direct measurement of gravity waves in September 2015, when the Advanced LIGO detector recorded a signal resulting from two massive stellar black holes spiralling together and merging 1,200,000,000 light years from Earth.

Detection

Passing gravitational waves will distort local spacetime, compressing and stretching distance. Compared to the other fundamental forces, gravity is incredibly weak, so the effects are minute. The LIGO discovery depended on sensing a change in distance of the order of 1 in 1,000,000,000,000,000,000,000. This is equivalent to the space between Earth and the centre of our Milky Way being stretched or compressed by 25 centimetres.

> The actual data recording the first direct measurement of a gravitational wave with the Advanced LIGO detector on 14 September 2015

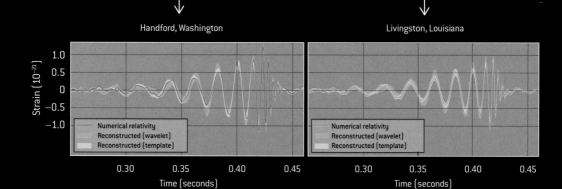

Handford, Washington

Livingston, Louisiana

Atomic theory

Chemical elements constitute all of the ordinary matter of the Universe, and atoms are the smallest units that have the chemical properties of pure elements. Ancient philosophies reasoned that matter was made up of discrete units, but it wasn't until the 1800s that science provided direct evidence. Since then, our understanding and description of atomic structure has continued to evolve.

The Thomson Model (1897)

Classical Model (460 BC–1890)

The idea that matter is made of indivisible units dates back to ancient philosophy, and early science established that compounds were composed of distincitive proportions of their constituent elements. Atoms were unique, indivisible and unalterable by chemical means.

Th

Extra spatial dimensions?

?

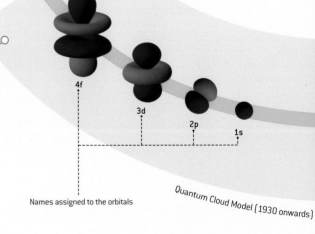

4f

3d

2p

1s

Along with particle-wave duality, a further implication of quantum theory is that there is a fundamental uncertainty of where an electron might be if we know its angular momentum, so for any energy level the position and, therefore, the orbital pattern of electrons can only be described in terms of the probability of finding the electron in one place or another. These probabilistic orbital clouds have specific shapes and sizes: spheres, rings and dumbbells in various orientations in space.

Names assigned to the orbitals

Quantum Cloud Model (1930 onwards)

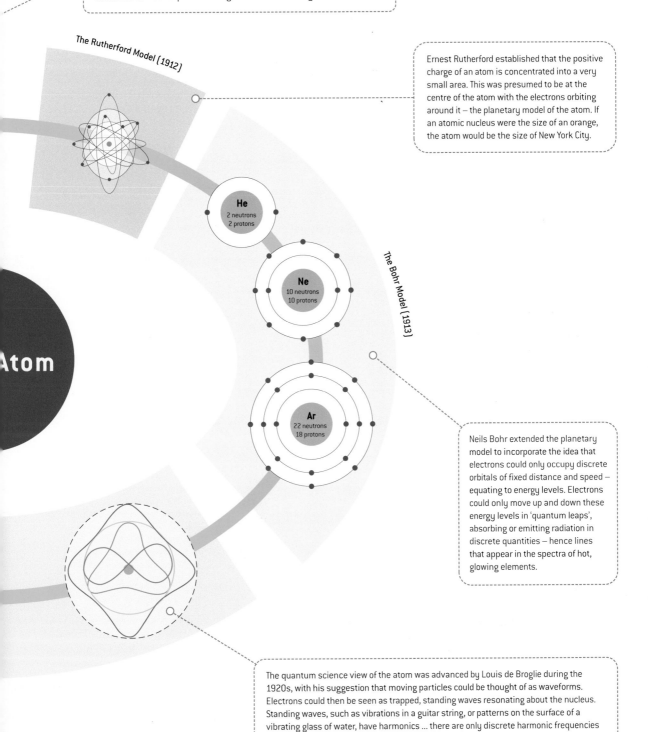

The discovery of the electron by J.J. Thomson in 1897 revealed that the atom has constituent parts. The electron is negatively charged and 1,800 times lighter than a hydrogen atom. The new model of the atom is described as a sea of positive charge embedded with negative electrons.

The Rutherford Model (1912)

Ernest Rutherford established that the positive charge of an atom is concentrated into a very small area. This was presumed to be at the centre of the atom with the electrons orbiting around it – the planetary model of the atom. If an atomic nucleus were the size of an orange, the atom would be the size of New York City.

The Bohr Model (1913)

He
2 neutrons
2 protons

Ne
10 neutrons
10 protons

Ar
22 neutrons
18 protons

Atom

Neils Bohr extended the planetary model to incorporate the idea that electrons could only occupy discrete orbitals of fixed distance and speed – equating to energy levels. Electrons could only move up and down these energy levels in 'quantum leaps', absorbing or emitting radiation in discrete quantities – hence lines that appear in the spectra of hot, glowing elements.

The quantum science view of the atom was advanced by Louis de Broglie during the 1920s, with his suggestion that moving particles could be thought of as waveforms. Electrons could then be seen as trapped, standing waves resonating about the nucleus. Standing waves, such as vibrations in a guitar string, or patterns on the surface of a vibrating glass of water, have harmonics ... there are only discrete harmonic frequencies that they can achieve. This was a way to describe the quantum energy states of electrons.

Quantum energy states

Quantum theory proposes that all subatomic particles behave, and can be thought of, as both points of matter and waveforms, tight little packets of waves. This is born out by the behaviour of electrons orbiting the atomic nucleus.

An excited electron is in a position to fall back down to a lower state of excitation – in doing so, it emits electromagnetic energy in the form of a photon ... light is emitted from the atom. This is why hot elements glow.

An atom is in its ground state at **absolute zero** – 0 kelvin (−273.15°C). As the temperature of an atom rises through the absorbtion of electromagnetic radiation (photons), thermal energy, the system becomes excited and achieves a higher energy state. This absorbtion of energy raises the energy state of the electrons orbiting the nucleus. Like a stone being rolled up a hill, the electrons gain in potential energy. But the discovery of **emission lines** indicated that electrons don't behave like this.

Hydrogen
(1 electron)

Helium
(2 electrons)

Mercury
(80 electrons)

Uranium
(92 electrons)

Excited

It's all about resonance!

The simplest 'absorption' of part of a photon's energy is when it scatters on an electron and gives up part of its energy to the electron, changing its wavelength in the process. The electron gains energy and increases its state of excitation ... it moves 'up the hill'.

Like an electron in an inner shell, a **valence electron** has the ability to absorb or release energy in the form of a photon. An energy gain can trigger an electron to jump to an outer shell – atomic excitation. Or the electron can even break free from its associated atom's valence shell, quiting the atom entirely – this is called **ionisation** and leads to the formation of a positive ion. When an electron loses energy – emitting a photon – then it can move to an inner shell which is not fully occupied.

Ground stat

EMISSION SPECTRUM

Spectral lines such as these are produced by splitting the light (electromagnetic radiation) emitted by a hot, glowing sample of a pure element. In the first half of the nineteenth century, experiments began to reveal that spectra weren't continuous and smooth: discrete, dark lines were seen in the sun's spectrum; corresponding brighter lines were spotted in the light from candles and arc-lamps. It wasn't until 1912 that Niels Bohr ushered in a new age of understanding and explained the lines in terms of particle/wave duality, resonance and quantum behaviour.

QUANTUM BEHAVIOUR

At the scale of subatomic particles matter behaves in a way that seems to defy common understanding. At this level all matter (including electrons) behaves both as if it's made of particles (familiar at greater scales ... a mote of dust, a grain of sand) and as if it is a wave ... a dense excitation of a field ... a wave packet.

Electrons, trapped by their attraction for the positively charged nucleus and orbiting at great speed, exhibit wave-like behaviour at quantum scales. They are trapped like the wave in a guitar string, or ripples on the surface of a pond. The electrons behave like a standing wave – and like the vibrations in a string – and settle in to a resonant frequency. They can only achieve discrete levels of vibration ... in this case quantum energy.

Electron orbitals depicted as standing waves ... they can only exist in energy states at distinct mathematical multiples ... just like musical harmonics.

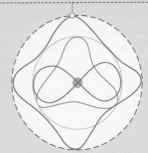

Star formation

Dying stars of great mass eventually collapse in on themselves and explode in supernovae, sending massive shockwaves and fusion remnants out into interstellar space. The shockwaves compress, heat and give momentum to molecular clouds of hydrogen and helium. In time ... a lot of time ... gravity causes the interstellar matter to clump and collapse, eventually forming the next generation of stars.

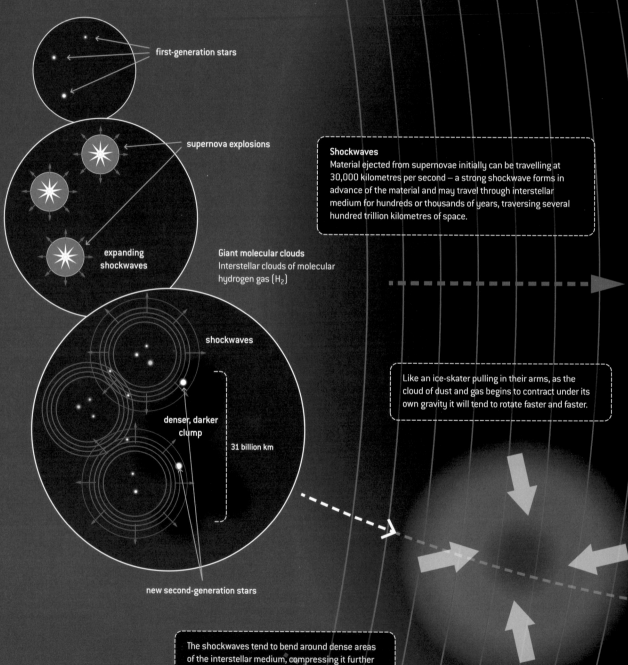

first-generation stars

supernova explosions

expanding shockwaves

Shockwaves
Material ejected from supernovae initially can be travelling at 30,000 kilometres per second – a strong shockwave forms in advance of the material and may travel through interstellar medium for hundreds or thousands of years, traversing several hundred trillion kilometres of space.

Giant molecular clouds
Interstellar clouds of molecular hydrogen gas (H_2)

shockwaves

Like an ice-skater pulling in their arms, as the cloud of dust and gas begins to contract under its own gravity it will tend to rotate faster and faster.

denser, darker clump

31 billion km

new second-generation stars

The shockwaves tend to bend around dense areas of the interstellar medium, compressing it further and imparting shearing and rotational forces.

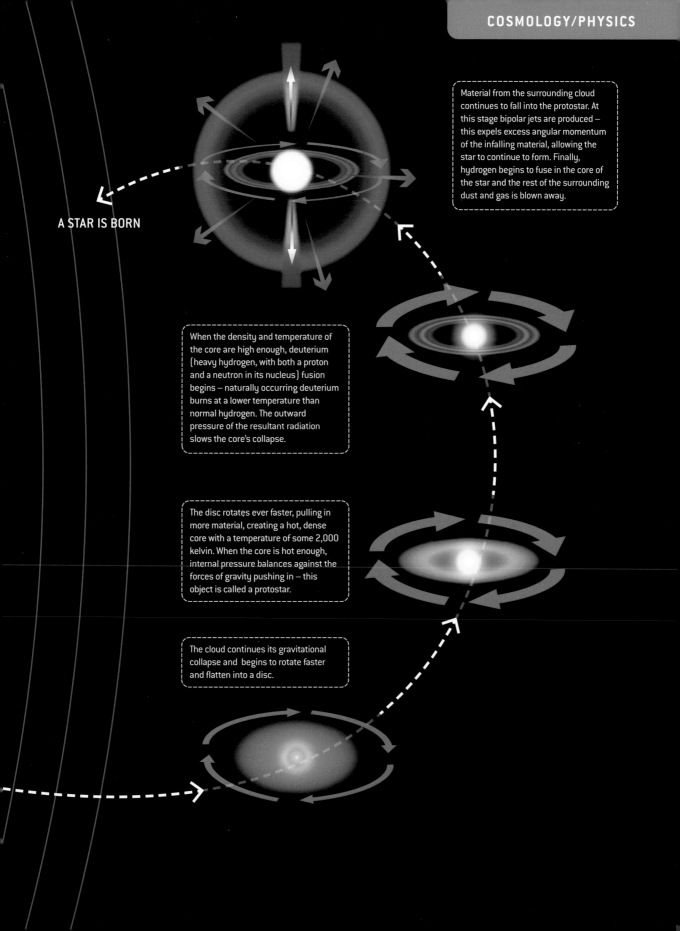

Material from the surrounding cloud continues to fall into the protostar. At this stage bipolar jets are produced – this expels excess angular momentum of the infalling material, allowing the star to continue to form. Finally, hydrogen begins to fuse in the core of the star and the rest of the surrounding dust and gas is blown away.

A STAR IS BORN

When the density and temperature of the core are high enough, deuterium (heavy hydrogen, with both a proton and a neutron in its nucleus) fusion begins – naturally occurring deuterium burns at a lower temperature than normal hydrogen. The outward pressure of the resultant radiation slows the core's collapse.

The disc rotates ever faster, pulling in more material, creating a hot, dense core with a temperature of some 2,000 kelvin. When the core is hot enough, internal pressure balances against the forces of gravity pushing in – this object is called a protostar.

The cloud continues its gravitational collapse and begins to rotate faster and flatten into a disc.

Stellar nucleosynthesis: the dawn of chemistry

All elements in the Universe heavier than hydrogen and helium – the calcium in your bones, the iron in your blood – were forged in the heart of ancient stars and dispersed in their death throes.

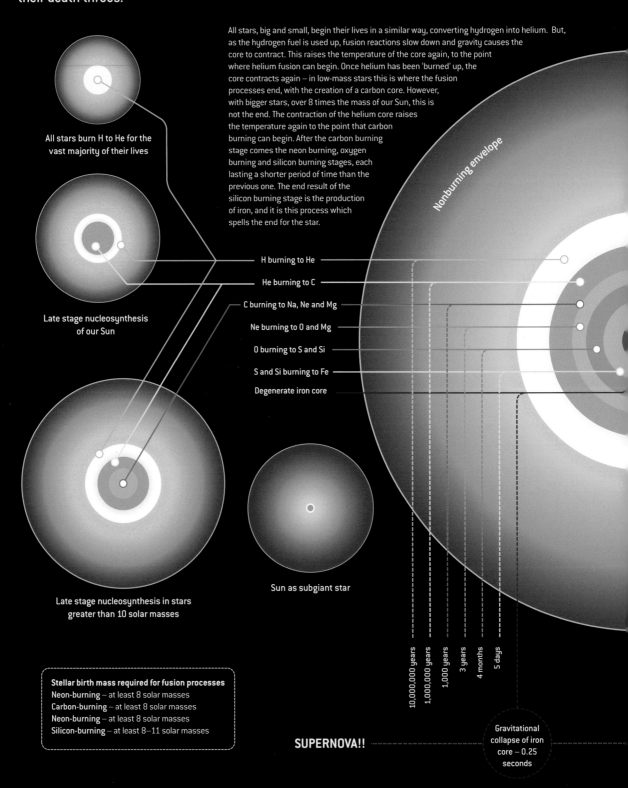

All stars, big and small, begin their lives in a similar way, converting hydrogen into helium. But, as the hydrogen fuel is used up, fusion reactions slow down and gravity causes the core to contract. This raises the temperature of the core again, to the point where helium fusion can begin. Once helium has been 'burned' up, the core contracts again – in low-mass stars this is where the fusion processes end, with the creation of a carbon core. However, with bigger stars, over 8 times the mass of our Sun, this is not the end. The contraction of the helium core raises the temperature again to the point that carbon burning can begin. After the carbon burning stage comes the neon burning, oxygen burning and silicon burning stages, each lasting a shorter period of time than the previous one. The end result of the silicon burning stage is the production of iron, and it is this process which spells the end for the star.

All stars burn H to He for the vast majority of their lives

Late stage nucleosynthesis of our Sun

Late stage nucleosynthesis in stars greater than 10 solar masses

Sun as subgiant star

Nonburning envelope

H burning to He

He burning to C

C burning to Na, Ne and Mg

Ne burning to O and Mg

O burning to S and Si

S and Si burning to Fe

Degenerate iron core

10,000,000 years
1,000,000 years
1,000 years
3 years
4 months
5 days

Stellar birth mass required for fusion processes
Neon-burning – at least 8 solar masses
Carbon-burning – at least 8 solar masses
Neon-burning – at least 8 solar masses
Silicon-burning – at least 8–11 solar masses

SUPERNOVA!!

Gravitational collapse of iron core – 0.25 seconds

Hydrogen burning

4–70,000,000 kelvin

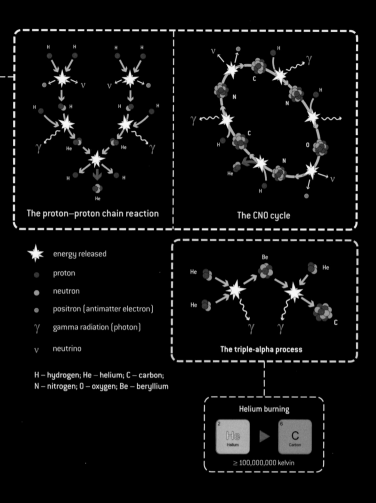

The proton–proton chain reaction

The CNO cycle

★ energy released

● proton

● neutron

● positron (antimatter electron)

γ gamma radiation (photon)

ν neutrino

H – hydrogen; He – helium; C – carbon;
N – nitrogen; O – oxygen; Be – beryllium

The triple-alpha process

Helium burning

≥ 100,000,000 kelvin

The structure of a massive star in its last throes of life. The timescales are calculated for a star of 25 solar masses

WHY IS IRON THE FINAL PRODUCT OF STELLAR FUSION?

Iron, with 26 protons, has the lowest potential nuclear energy of all the elements: energy is released in the fusion of lighter elements into iron. Energy is also released when you split heavier elements back into iron. It is the final destination in the heart of a star ...

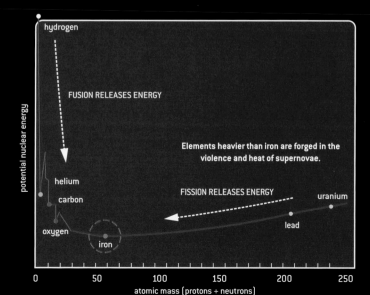

hydrogen

FUSION RELEASES ENERGY

Elements heavier than iron are forged in the violence and heat of supernovae.

FISSION RELEASES ENERGY

helium

carbon

oxygen

iron

lead

uranium

potential nuclear energy

atomic mass (protons + neutrons)

Supernovae create, fuse and eject the bulk of the chemical elements produced by nucleosynthesis. Supernovae play a significant role in enriching the interstellar medium with **higher mass elements**.

The elements

The atomic nucleus consists of a number of protons – this number is called the **atomic number**; and a number of neutrons, referred to as the **neutron number**. As neutrons are positively charged, they are matched by a equal number of negatively charged electrons in nested orbital shells around the nucleus. The electrons of one atom can interact with the electrons of another – this is the basis of chemical bonding. Thus the atomic number of an element determines its chemical properties.

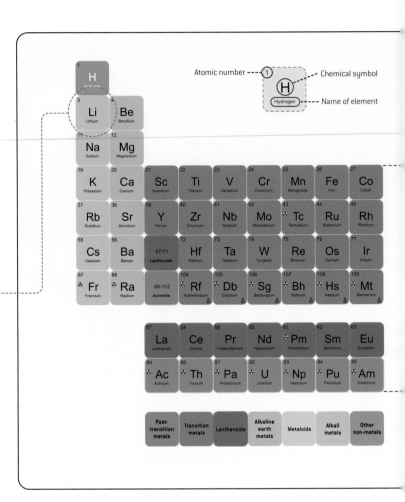

Atomic number · Chemical symbol · Name of element

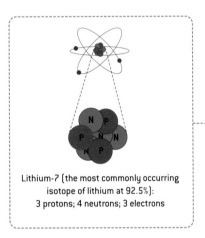

Lithium-7 (the most commonly occurring isotope of lithium at 92.5%):
3 protons; 4 neutrons; 3 electrons

Post-transition metals · Transition metals · Lanthanoids · Alkaline earth metals · Metalloids · Alkali metals · Other non-metals

Relative mass abundance in Earth's bulk

ORIGINS OF THE ELEMENTS

Big Bang

Cosmic rays

Small stars/Big stars

Big stars

Big stars/Supernovae

Supernovae

Manmade

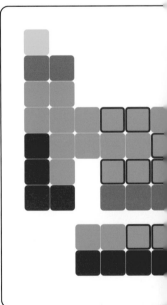

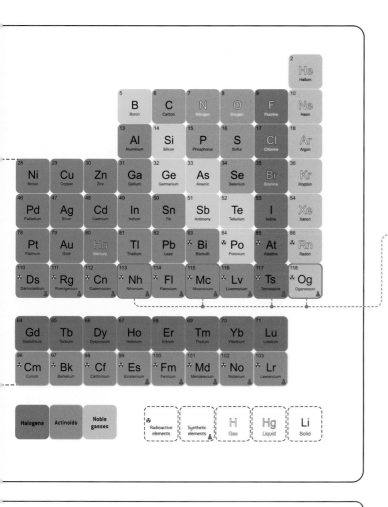

In the galaxy as a whole, hydrogen and helium – created between 10 seconds and 20 minutes after the Big Bang – still predominate. The heavier elements, created in the furnace of the stars and released into interstellar space as those stars died, go on to form new protostars and planetary discs. Our Earth is rich in iron, magnesium, silicon and oxygen forged in other star systems, hugely distant in time and interstellar distance.

Life, evolved in the medium of H_2O, has harnessed the ability to extract hydrogen, oxygen and nitrogen from the atmosphere and hydrosphere. We are air and water, with a sprinkling of dirt … all wrought from stardust.

In December 2015 four new synthetic elements earned a spot on the periodic table, completing the seventh row. They were named in June 2016.

Relative mass abundance in our Galaxy

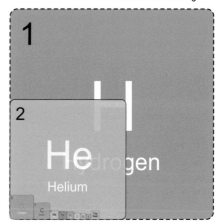

Relative mass abundance in Human body

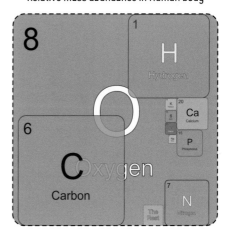

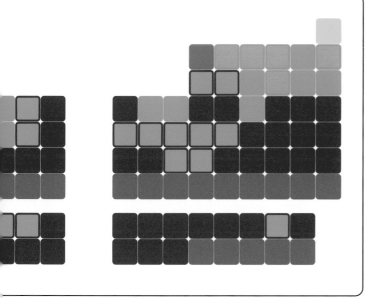

Chemical reactivity

The chemical reactivity of the elements — the tendency and violence with which they will react and combine with other elements — and the kind of elements they will want to react with, is determined by the structure and size of their electron orbitals.

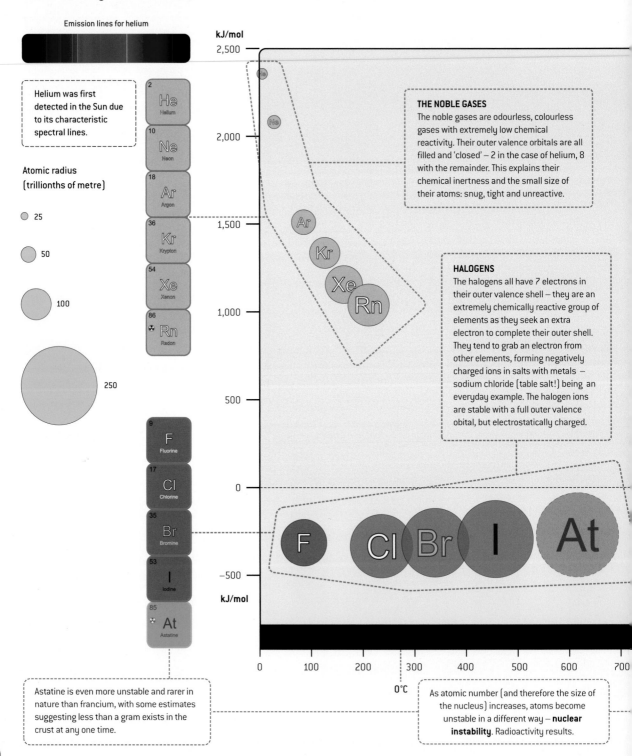

Emission lines for helium

Helium was first detected in the Sun due to its characteristic spectral lines.

Atomic radius
(trillionths of metre)

- 25
- 50
- 100
- 250

THE NOBLE GASES
The noble gases are odourless, colourless gases with extremely low chemical reactivity. Their outer valence orbitals are all filled and 'closed' — 2 in the case of helium, 8 with the remainder. This explains their chemical inertness and the small size of their atoms: snug, tight and unreactive.

HALOGENS
The halogens all have 7 electrons in their outer valence shell — they are an extremely chemically reactive group of elements as they seek an extra electron to complete their outer shell. They tend to grab an electron from other elements, forming negatively charged ions in salts with metals — sodium chloride (table salt!) being an everyday example. The halogen ions are stable with a full outer valence obital, but electrostatically charged.

Astatine is even more unstable and rarer in nature than francium, with some estimates suggesting less than a gram exists in the crust at any one time.

As atomic number (and therefore the size of the nucleus) increases, atoms become unstable in a different way — **nuclear instability**. Radioactivity results.

Some elements long to acquire electrons; others would love to lose them. This tendency is dependent on the number of electrons in an element's outermost orbital (the **valence electrons**) and how far from the nucleus that orbital is. Atoms are most stable if they have a filled or empty outer layer of electrons – the filled outer orbital will generally be filled by 8 electrons. Atoms will gain or lose electrons to become ions, or share electrons with other atoms to become covalent compounds.

Emission lines for lithium

Lithium is the lightest metal and the least dense solid element. As well as being very chemically reactive, its nucleus exists on the verge of instability, having the lowest nuclear binding energy of any atom. This property accounts for its relative scarcity (26th most common) in our Solar System, despite its low atomic weight.

Ionisaton energy
(energy required to remove an electron)

ALKALI METALS
The alkali metals all have a single electron in their outer valence shell – they are all highly reactive, as they seek to lose their lone electron. They are the flipside of the halogens and tend to form positive ions in salts. They have the largest atomic diameters, as their outer electrons are the least 'closed' and the most likely to be orbiting far from the nucleus.

Electron affinity
(energy released when electron acquired)

Francium is very unstable, existing in nature only as part of natural radioactive decay – with a half-life of 22 minutes, it is estimated that there are only ever 20–30 grams (an ounce) of naturally occurring francium in all the world's crust at any one moment.

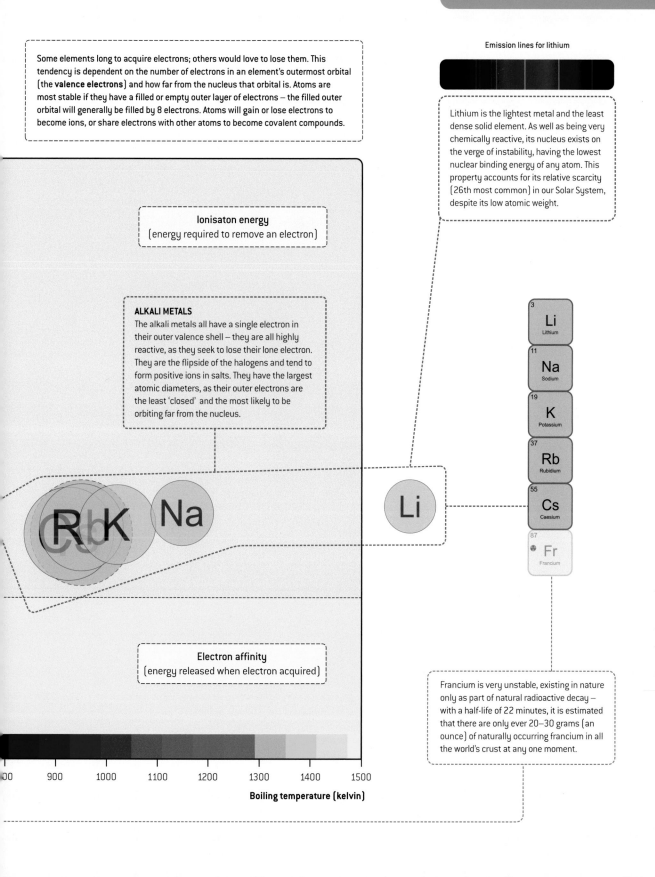

| 3 Li Lithium |
| 11 Na Sodium |
| 19 K Potassium |
| 37 Rb Rubidium |
| 55 Cs Caesium |
| 87 Fr Francium |

| 00 | 900 | 1000 | 1100 | 1200 | 1300 | 1400 | 1500 |

Boiling temperature (kelvin)

Metals

Metallic elements are generally defined by physical characteristics, principally their ability to conduct electricity and heat. In addition, they are typically hard solids that are shiny, can be melted, hammered into sheets and drawn into wires.

EXOTIC METALS

NEODYMIUM
Melting point: 1024 °C / Density: 89% of iron
A component in the strong magnets used in small speakers, hard drives and hybrid cars. Neodymium magnets can lift thousands of times their own weight.

EUROPIUM
Melting point: 826 °C / Density: 67% of iron
Used in the printing of euro banknotes – it glows red under UV light, and forgeries can be detected by the lack of this red glow. Europium is one of the least abundant elements – accounting for only about 0.00000005 % of the entire Universe.

GALLIUM
Melting point: 30 °C / Density: 75% of iron
Prized for the use of gallium compounds as semiconductors in low-power chips used in mobile phones and high-efficiency solar cells used in satellites and space robots. Melting at a little below blood heat, solid gallium will liquify in the palm of the hand.

TERBIUM
Melting point: 1356 °C / Density: 105% of iron
Used in the printing of euro banknotes – it glows green under UV light.

TUNGSTEN
Melting point: 3422°C / Density: 245% of iron
The highest melting point of any metal – only diamond has a higher melting point, though it doesn't really melt as it **sublimates** ... turns from solid to gas directly.

Sitting at the top of the alkali metals in the periodic table, hydrogen is obviously not, under ordinary conditions, a metal. However, it is predicted that under an immense pressure of around 250,000 atmospheres, hydrogen atoms would display metallic properties, losing hold over their electrons. It is thought that the gravitationally compressed interiors of Jupiter and Saturn contain huge quantites of metallic hydrogen.

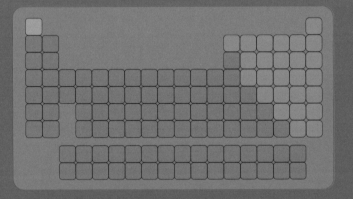

hydrogen

non-metals

metals

ATMOSPHERE (90 kilometres)

EARTH
Total mass: 5.92 x 10^{24} kg
Diameter : 12,742 km

Metals can be seen as a lattice of positive **ions** glued together by a 'sea' of shared valence electrons. Metals conduct electricity because the shared electrons are free to move about – electrons moving from one part of the metal will cause electrons from surrounding areas to rush in to replace them. Metals can be drawn into wire or pounded into sheets because the metal ions can slide past each other but still be bound together by the shared electrons.

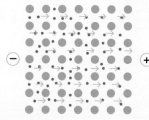

Applying a voltage will cause the sea of electrons to flow towards the positive charge.

Physical force deforms the lattice of positive ions, which are able to slide past each other.

IRON
Total mass: 1.9×10^{24} kg
Diameter of solid sphere: 7,738 km

COPPER
Total mass: 3.6×10^{20} kg
Diameter of solid sphere: 425 km

ALUMINIUM
Total mass: 9.5×10^{22} kg
Diameter of solid sphere: 4,054 km

GOLD
Total mass: 9.6×10^{17} kg
Diameter of solid sphere: 46 km

PLATINUM
Total mass: 1.1×10^{19} kg
Diameter of solid sphere: 100 km

URANIUM
Total mass: 1.2×10^{17} kg
Diameter of solid sphere: 23 km

ESTIMATED TOTAL QUANTITIES OF METAL ELEMENTS IN THE EARTH

Carbon

Life is based on carbon – an extraordinarily versatile element because of its ability and tendency to form four covalent bonds. This allows it to form a huge variety of polymerised shapes, which form the backbone of the molecules that form living matter.

6
C
Carbon

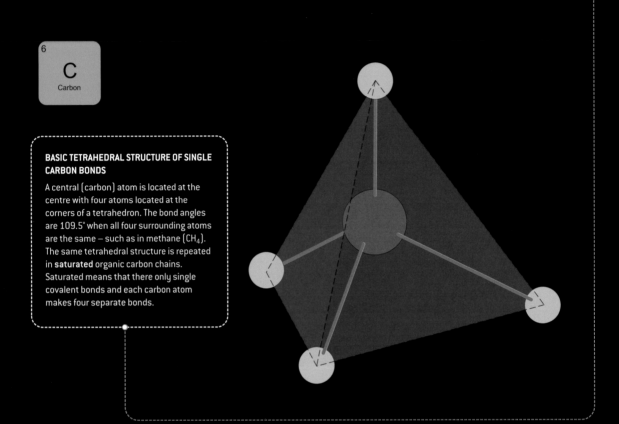

BASIC TETRAHEDRAL STRUCTURE OF SINGLE CARBON BONDS

A central (carbon) atom is located at the centre with four atoms located at the corners of a tetrahedron. The bond angles are 109.5° when all four surrounding atoms are the same – such as in methane (CH_4). The same tetrahedral structure is repeated in **saturated** organic carbon chains. Saturated means that there only single covalent bonds and each carbon atom makes four separate bonds.

DOUBLE COVALENT BOND

Carbon atoms readily form double covalent bonds, where two pairs of electrons are shared. Whereas single bonds are free to rotate, double bonds are fixed and lend rigidity to carbon chains: they are present in unsaturated oils and fats.

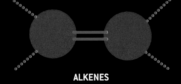

ALKENES

TRIPLE COVALENT BOND

Triple bonds, the rarest form of carbon bond, are also the strongest. Over a thousand naturally occurring alkynes have been discovered, many are biotoxins.

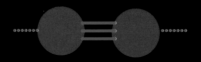

ALKYNES

COVALENT BONDING

The strongest kind of chemical bonding that involves the sharing of electron pairs between atoms. The stable balance of attractive and repulsive forces between atoms, when they share electrons, is known as covalent bonding. In most cases the sharing of electrons allows each atom to attain the equivalent of a full outer shell of electrons.

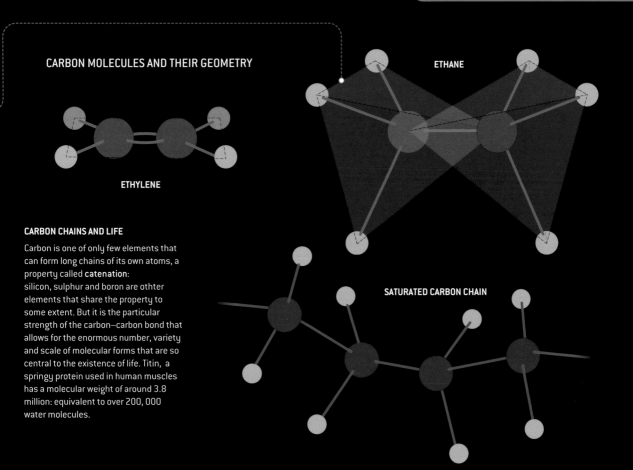

CARBON MOLECULES AND THEIR GEOMETRY

ETHYLENE

ETHANE

SATURATED CARBON CHAIN

CARBON CHAINS AND LIFE

Carbon is one of only few elements that can form long chains of its own atoms, a property called **catenation**: silicon, sulphur and boron are othter elements that share the property to some extent. But it is the particular strength of the carbon–carbon bond that allows for the enormous number, variety and scale of molecular forms that are so central to the existence of life. Titin, a springy protein used in human muscles has a molecular weight of around 3.8 million: equivalent to over 200, 000 water molecules.

ALLOTROPES OF CARBON

Carbon can exist in a variety of different structural forms known as **allotropes**, where the are bonded together in a different way. The allotropes of carbon include:

- **amorphous carbon** – does not have any crystalline structure; present in soot and charcoal
- **diamond** – atoms are bonded together in a tetrahedral lattice arrangement
- **graphite** – atoms are bonded together in sheets of a hexagonal lattice
- **graphene** – single sheets of graphite
- **fullerenes** – atoms are bonded together in spherical, tubular or ellipsoidal structures

Diamond

Amorphous carbon (charcoal)

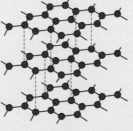

Graphite

FULLERENES

The first fullerene molecule to be discovered (1985) was **buckminsterfullerene**, named in homage to Buckminster Fuller, whose geodesic domes the stucture resembles. Fullerenes have been found to occur in nature, most recently they have been detected in outer space, possibly providing a cosmic source for the seeds of life on Earth.

Buckminsterfullerene C_{60}

carbon nanotube

States of matter

Despite the plethora elements and compounds that compose the Universe, there are only four states of matter: only four conditions that describe how the atoms or molecules that comprise that substance behave and how they interact with their physical environment.

H
At room temperature and pressure, hydrogen is a colourless, odourless, tasteless, non-toxic, nonmetallic, highly combustible diatomic gas that becomes solid below 14 K. Metallic hydrogen is a theoretical exotic phase of hydrogen in which it behaves like an electrical conductor. At high pressure, perhaps 5,000 times atmospheric pressure, hydrogen might exist as a conducting liquid rather than a gas. Liquid metallic hydrogen is thought to be present in large amounts in the gravitationally compressed interiors of Jupiter and Saturn.

Nb/Mo/Ta/W/Re/Os
Refractory metals are a class of metals that are extraordinarily resistant to heat and wear. They are chemically inert and have a relatively high density. Osmium is the densest naturally occurring element ... twice as dense as lead. Tungsten has the highest melting point of any element other than carbon.

EXOTIC STATES OF MATTER

These four states of matter are what are observable in everyday life. Other states are known to exist in extreme situations, such as **Bose–Einstein condensates, neutron-degenerate matter** and **quark–gluon plasma**, which arise in situations of extreme cold, extreme density and extremely high-energy matter respectively. Some other states are believed to be possible but remain theoretical.

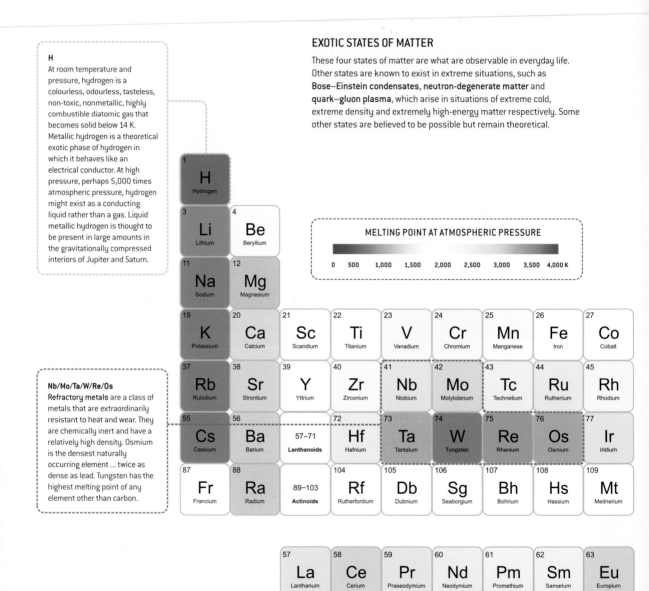

MELTING POINT AT ATMOSPHERIC PRESSURE

0   500   1,000   1,500   2,000   2,500   3,000   3,500   4,000 K

PHASE TRANSITIONS

Changes between states of matter are reversible: as temperature and pressure are varied, a substance can make transitions to and from each of the states — though not all compounds can exist in all states.

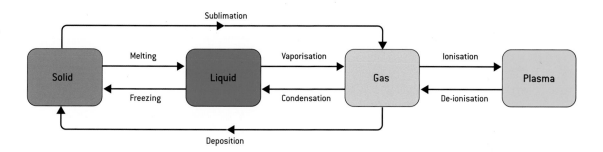

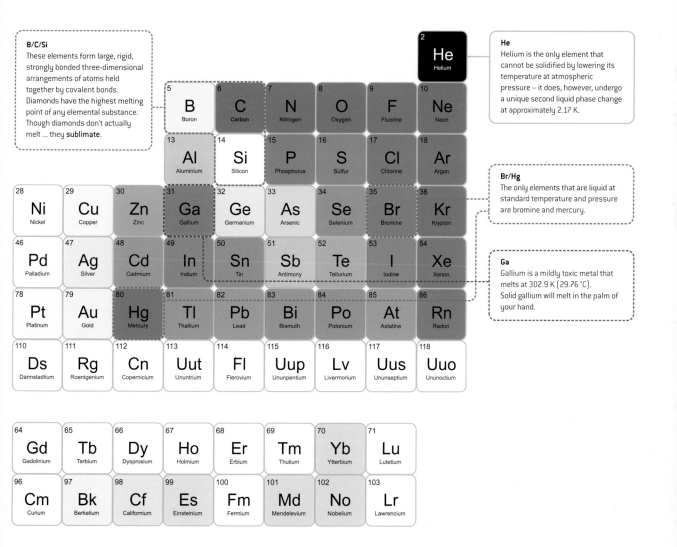

B/C/Si
These elements form large, rigid, strongly bonded three-dimensional arrangements of atoms held together by covalent bonds. Diamonds have the highest melting point of any elemental substance. Though diamonds don't actually melt … they **sublimate**.

He
Helium is the only element that cannot be solidified by lowering its temperature at atmospheric pressure — it does, however, undergo a unique second liquid phase change at approximately 2.17 K.

Br/Hg
The only elements that are liquid at standard temperature and pressure are bromine and mercury.

Ga
Gallium is a mildly toxic metal that melts at 302.9 K (29.76 °C). Solid gallium will melt in the palm of your hand.

Supernovae

Large stars burn brightly and die quickly, their lives spanning millions, rather than billions, of years. Stars of more than three times the mass of our Sun end their lives explosively as supernovae. These cosmic outbursts forge most of the heavier nuclei that occur in nature.

TYPE II SUPERNOVA

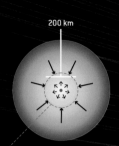

200 km

Iron nuclei neither fuse nor break apart, so the iron core of the star ceases nuclear reactions: no longer generating heat, the core pressure drops and the overlying material suddenly rushes in, crushing the core with vast gravitational force. After millions of years of life the core collapse takes less than half a second.

As a massive star nears the end of its life, it evolves an onion-skin structure as heavier elements are progressively fused towards the centre. By this point the star will be hundreds of millions of kilometres in diameter – only the inner core is shown here.

Neutron-rich core

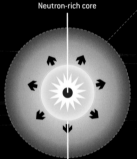

The infalling mass of the star crushes the core to nuclear densities, overshoots and rebounds, setting up an outward-travelling pressure wave.

Supernova explosions can initally outshine the combined output of their home galaxies and do so for some weeks.

Pressure wave
30–40,000 kilometres per second

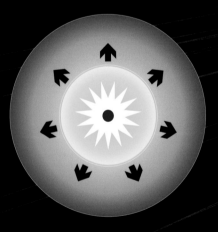

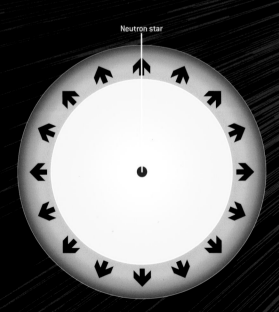

Neutron star

Neutrinos, pouring out of the developing neutron star, force the pressure wave outward unevenly. The shockwave rushes through the entire star, blowing it apart.

ELEMENTS EXCLUSIVELY SYNTHESISED IN SUPERNOVAE

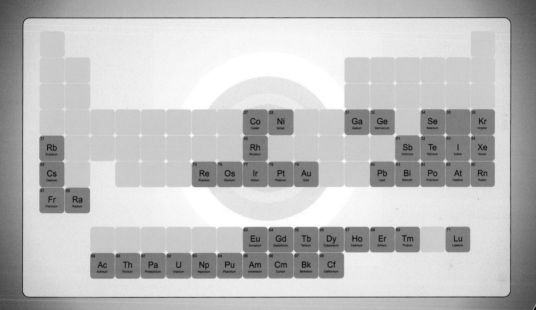

Space is not a vacuum, and is better described as **interstellar medium**. The interstellar medium is the matter that exists in the space between the star systems in a galaxy. This includes gas in ionic, atomic, and molecular form, as well as dust and cosmic rays (fast moving, energetic nuclear particles). The interstellar medium is composed primarily of hydrogen followed by helium with trace amounts of carbon, oxygen and nitrogen. A supernova expels much or all of a star's material at a velocity of up to **30,000 kilometres per second** – 10% of the speed of light – driving a shockwave into the surrounding interstellar medium. Supernova shockwaves release a huge amount of particles and ions outward, which travel through the space and interact with local interstellar molecular clouds. It was a shock wave from a nearby supernova that resulted in the collapse of a molecular cloud that formed our Solar System.

Radioactive elements

All elements can exist in variant forms (called **isotopes**), with a smaller or larger number of neutrons in their nucleus. These different nuclei are called **nuclides**. There is an optimum balance of neutrons to protons, resulting in a stable nucleus: other combinations are inherently unstable, and beyond a certain mass all nuclei are unstable and radioactive.

Unstable elements, and unstable isotopes of stable elements, decompose through a process called radioactive decay. This decay can be extremely fast, extremely slow ... and all points in between. The rate of decay is described by the length of time it takes half of the atoms in a sample to decay – the **half-life**. The protons making up the nucleus are positively charged and repel each other. The neutrons attract themselves and the protons – they act as 'glue'. The more protons in the nucleus, the heavier the element, and the more neutrons that are needed to try and keep all the protons together.

PRIMORDIAL NUCLIDES
There are 288 known nuclides that have existed since before the formation of the Earth: this includes the 254 stable nuclides, isotopes of 80 elements, plus another 34 nuclides that have half-lives long enough to have survived from the formation of the Earth.

Isotopes/nuclides are identified by their atomic weight, which is the number of protons, plus the nuber of neutrons: rubidium 87 has 37 protons and 50 neutrons in each atomic nucleus.

Promethium is one of two elements lighter than lead with no stable isotopes.

Tellurium 128 (half-life: 2.2×10^{24} years) has the longest half-life among all radionuclides, which is approximately 160 trillion times the age of the known Universe.

Rubidium 87 (half-life: 49 billion years).

Iron 56 is the most stable nucleus in the Universe.

Carbon 14 is a naturally occurring radioisotope of carbon with a half-life of 5,730 years. In living organisms, the proportion of carbon 14 is the same as the surrounding environment. After death the levels slowly reduce, as the carbon 14 decays. This is the basis for carbon dating, which can be used to date archeological organic materials up to 60,000 years old.

Number of protons

Number of neutrons

EUREKA! AN INFOGRAPHIC GUIDE TO SCIENCE

ISLAND OF STABILITY
The stability of nuclides is high if the total number of protons and neutrons adds up to a number that allows packing in complete shells within the atomic nucleus: these are known as **magic numbers**, the most recognised being 2, 8, 20, 28, 50, 82 and 126. It is predicted that there could be a class of extremely heavy nuclides with enhanced stability – but making these synthetic nuclides proves an allusive art.

- Unbihexium 310 ?
- ? • Unbinilium 304 ?
- Flerovium 298 ?

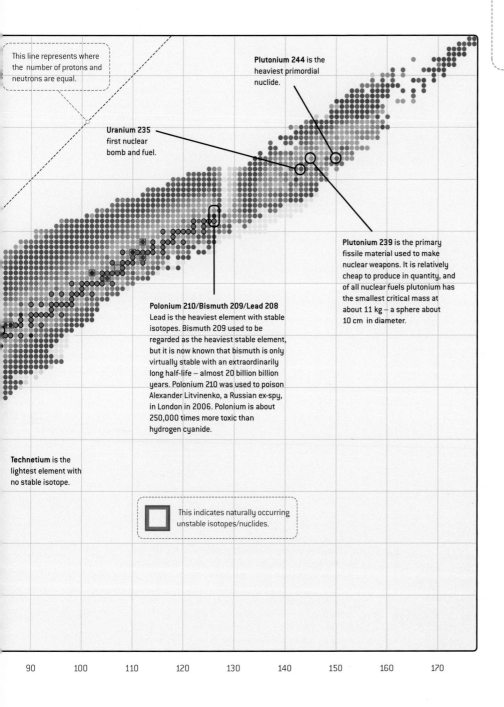

This line represents where the number of protons and neutrons are equal.

Plutonium 244 is the heaviest primordial nuclide.

Uranium 235
first nuclear bomb and fuel.

Plutonium 239 is the primary fissile material used to make nuclear weapons. It is relatively cheap to produce in quantity, and of all nuclear fuels plutonium has the smallest critical mass at about 11 kg – a sphere about 10 cm in diameter.

Polonium 210/Bismuth 209/Lead 208
Lead is the heaviest element with stable isotopes. Bismuth 209 used to be regarded as the heaviest stable element, but it is now known that bismuth is only virtually stable with an extraordinarily long half-life – almost 20 billion billion years. Polonium 210 was used to poison Alexander Litvinenko, a Russian ex-spy, in London in 2006. Polonium is about 250,000 times more toxic than hydrogen cyanide.

Technetium is the lightest element with no stable isotope.

This indicates naturally occurring unstable isotopes/nuclides.

NUCLIDE HALF-LIFE

○ Stable

10^{14} years
10^{12} years
10^{10} years
10^{8} years
10^{6} years
10^{4} years
100 years
1 year
10^{6} seconds
10^{4} seconds
100 seconds
1 second
10^{-2} seconds
10^{-4} seconds
10^{-6} seconds
10^{-8} seconds

Unstable/Unknown

90   100   110   120   130   140   150   160   170

Nuclear fission

Lead is the heaviest stable element; all elements with heavier nuclei are unstable and sooner or later will break apart into smaller nuclei, releasing fast moving particles such as neutrons and protons, losing a little mass in the process. The lost mass becomes energy multiplying by the speed of light squared.

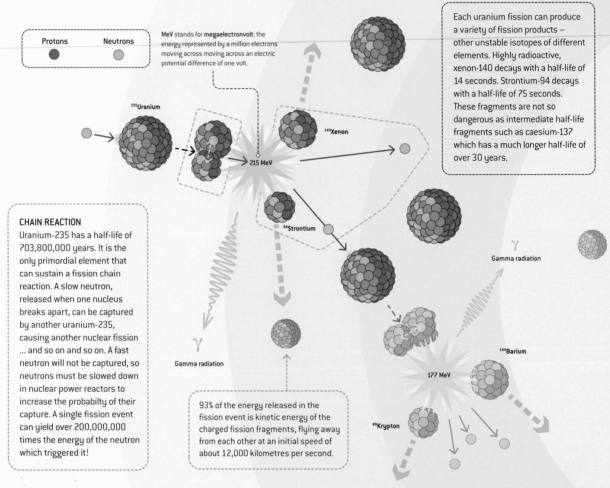

Protons   Neutrons

MeV stands for **megaelectronvolt**: the energy represented by a million electrons moving across moving across an electric potential difference of one volt.

^{235}Uranium

215 MeV

^{140}Xenon

^{94}Strontium

Gamma radiation

^{177}MeV

^{144}Barium

Gamma radiation

^{89}Krypton

Each uranium fission can produce a variety of fission products – other unstable isotopes of different elements. Highly radioactive, xenon-140 decays with a half-life of 14 seconds. Strontium-94 decays with a half-life of 75 seconds. These fragments are not so dangerous as intermediate half-life fragments such as caesium-137 which has a much longer half-life of over 30 years.

CHAIN REACTION

Uranium-235 has a half-life of 703,800,000 years. It is the only primordial element that can sustain a fission chain reaction. A slow neutron, released when one nucleus breaks apart, can be captured by another uranium-235, causing another nuclear fission ... and so on and so on. A fast neutron will not be captured, so neutrons must be slowed down in nuclear power reactors to increase the probabilty of their capture. A single fission event can yield over 200,000,000 times the energy of the neutron which triggered it!

93% of the energy released in the fission event is kinetic energy of the charged fission fragments, flying away from each other at an initial speed of about 12,000 kilometres per second.

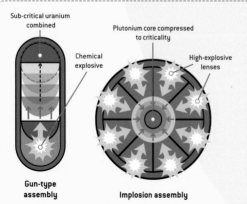

Sub-critical uranium combined

Chemical explosive

Plutonium core compressed to criticality

High-explosive lenses

Gun-type assembly

Implosion assembly

NUCLEAR BOMBS

All existing nuclear warheads derive some or all of their explosive energy from nuclear fission reactions. Uranium or plutonium is assembled into a supercritical mass – the amount of material needed to start a chain reaction – either by shooting one piece of sub-critical material into another or by compressing a hollow, sub-critical sphere of material using chemical explosives to many times its original density (the **implosion** method). The challenge with nuclear weapons design is to ensure that a significant fraction of the fissile material is consumed before the weapon destroys itself. The bomb dropped on Hiroshima contained 64 kg of enriched uranium, of which less than a kilogram underwent nuclear fission.

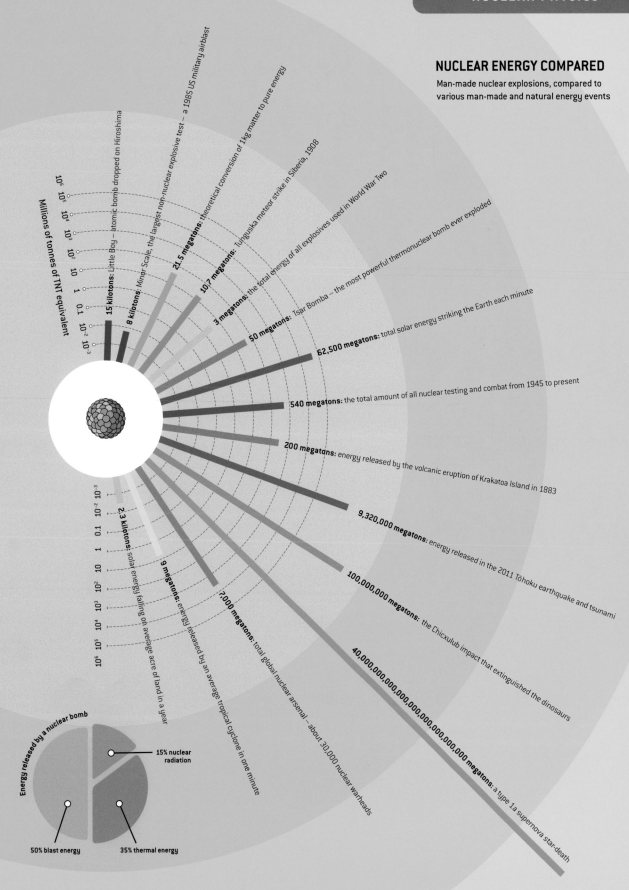

NUCLEAR ENERGY COMPARED

Man-made nuclear explosions, compared to
various man-made and natural energy events

Millions of tonnes of TNT equivalent

10^6 10^5 10^4 10^3 10^2 10 1 0.1 10^{-2} 10^{-3}

15 kilotons: Little Boy – atomic bomb dropped on Hiroshima

8 kilotons: Minor Scale, the largest non-nuclear explosive test – a 1985 US military airblast

21.5 megatons: theoretical conversion of 1kg matter to pure energy

10.7 megatons: Tunguska meteor strike in Siberia, 1908

3 megatons: the total energy of all explosives used in World War Two

50 megatons: Tsar Bomba – the most powerful thermonuclear bomb ever exploded

62,500 megatons: total solar energy striking the Earth each minute

540 megatons: the total amount of all nuclear testing and combat from 1945 to present

200 megatons: energy released by the volcanic eruption of Krakatoa Island in 1883

9,320,000 megatons: energy released in the 2011 Tōhoku earthquake and tsunami

100,000,000 megatons: the Chicxulub impact that extinguished the dinosaurs

40,000,000,000,000,000,000,000,000,000 megatons: a type 1a supernova star-death

7,000 megatons: total global nuclear arsenal – about 30,000 nuclear warheads

9 megatons: energy released by an average tropical cyclone in one minute

2.3 kilotons: solar energy falling on average acre of land in a year

10^{-3} 10^{-2} 0.1 1 10 10^2 10^3 10^4 10^5 10^6

Energy released by a nuclear bomb

15% nuclear radiation

50% blast energy

35% thermal energy

Star lifecycles

Our solar system coalesced about 4.6 billion years ago from molecular clouds produced by a previous generation of exploding stars. Meanwhile, our Sun is busy making its own slightly heavier elements, but those aren't ours ... they will belong to the next 'generation' of solar systems, born after our Sun dies. The life of a star, its length and ending, is determined by its starting mass.

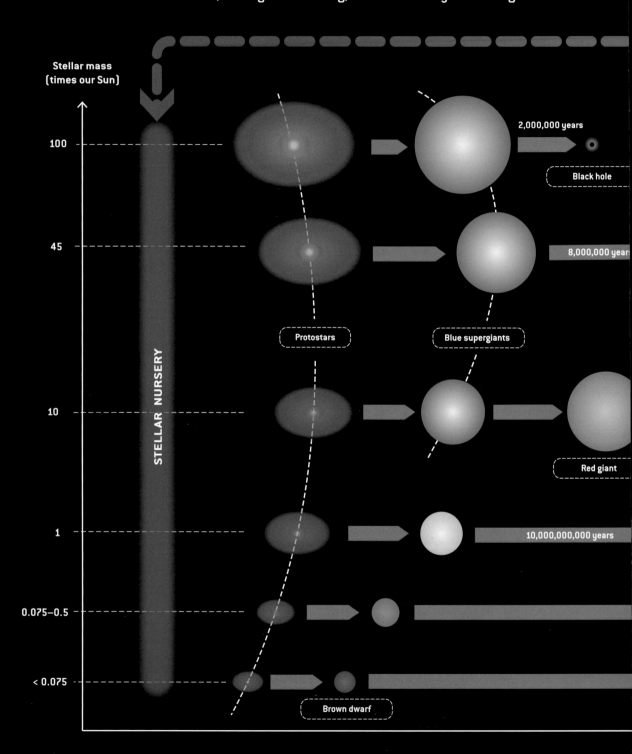

Stellar mass (times our Sun)

100

45

10

1

0.075–0.5

< 0.075

STELLAR NURSERY

Protostars

Blue supergiants

Red giant

Brown dwarf

2,000,000 years

Black hole

8,000,000 years

10,000,000,000 years

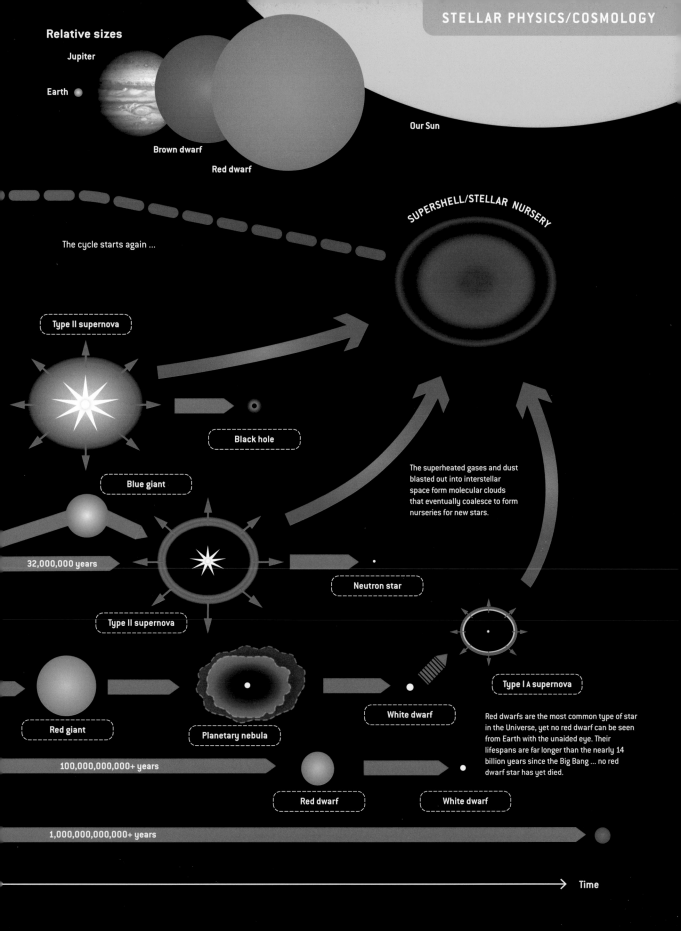

Relative sizes

Jupiter

Earth

Brown dwarf

Red dwarf

Our Sun

SUPERSHELL/STELLAR NURSERY

The cycle starts again ...

Type II supernova

Black hole

Blue giant

The superheated gases and dust blasted out into interstellar space form molecular clouds that eventually coalesce to form nurseries for new stars.

32,000,000 years

Neutron star

Type II supernova

Type I A supernova

Red giant

Planetary nebula

White dwarf

Red dwarfs are the most common type of star in the Universe, yet no red dwarf can be seen from Earth with the unaided eye. Their lifespans are far longer than the nearly 14 billion years since the Big Bang ... no red dwarf star has yet died.

100,000,000,000+ years

Red dwarf

White dwarf

1,000,000,000,000+ years

Time

Galaxies

Roughly one hundred billion galaxies are scattered throughout our observable Universe, each a glorious and complex system that can contain hundreds of billions of stars.

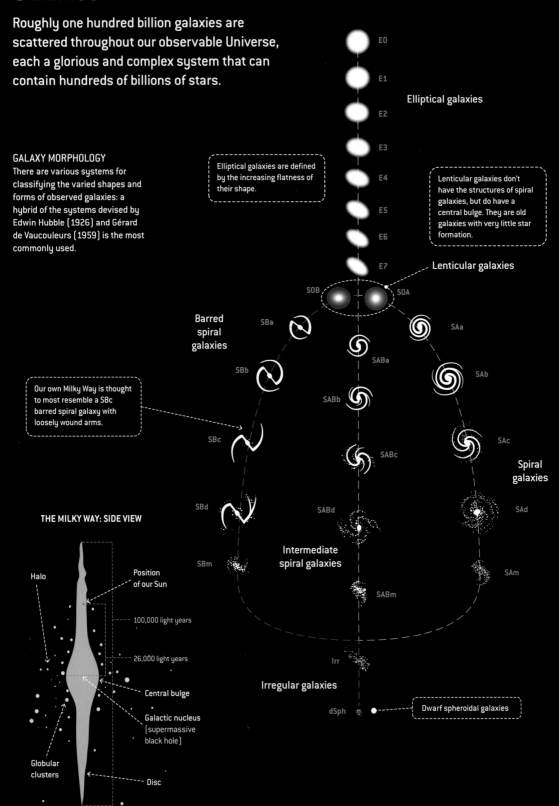

GALAXY MORPHOLOGY
There are various systems for classifying the varied shapes and forms of observed galaxies: a hybrid of the systems devised by Edwin Hubble (1926) and Gérard de Vaucouleurs (1959) is the most commonly used.

E0

E1

E2

E3

E4

E5

E6

E7

Elliptical galaxies

Elliptical galaxies are defined by the increasing flatness of their shape.

Lenticular galaxies don't have the structures of spiral galaxies, but do have a central bulge. They are old galaxies with very little star formation.

Lenticular galaxies

SOB   SOA

Barred spiral galaxies

SBa

SBb

Our own Milky Way is thought to most resemble a SBc barred spiral galaxy with loosely wound arms.

SBc

SBd

SBm

SABa

SABb

SABc

SABd

SABm

Intermediate spiral galaxies

SAa

SAb

SAc

SAd

SAm

Spiral galaxies

Irr

Irregular galaxies

dSph

Dwarf spheroidal galaxies

THE MILKY WAY: SIDE VIEW

Halo

Position of our Sun

100,000 light years

26,000 light years

Central bulge

Galactic nucleus (supermassive black hole)

Globular clusters

Disc

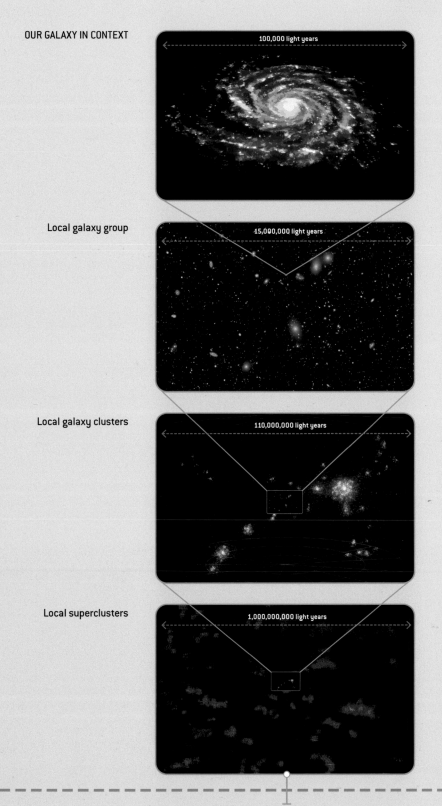

OUR GALAXY IN CONTEXT

100,000 light years

Local galaxy group

15,000,000 light years

Local galaxy clusters

110,000,000 light years

Local superclusters

1,000,000,000 light years

At this scale the observable Universe would be 6.5 metres across

Observable Universe: 93,000,000,000 light years

Black holes

Einstein's field equations, the fiendishly complex maths that form the foundation of his theory of general relativity, predicted the existence of black holes. Concentrations of matter so intense, that to escape their gravitational clutches, you have to travel faster than the speed of light ... a Universal impossibilty. There is no escape.

PRIMORDIAL BLACK HOLES

Small masses would have extremely small Schwarzschild radii. A mass similar to that of Mount Everest would have to be compressed into a space much smaller than a nanometre, to create a black hole. No known mechanism could create such extremely compact objects. However, it is possible that conditions shortly after the Big Bang, when densities were extremely high, might have allowed such microscopic black holes to form. Therefore these hypothetical miniature black holes are called primordial black holes.

BLACK HOLE MK I:
Non-spinning Schwarzschild black hole

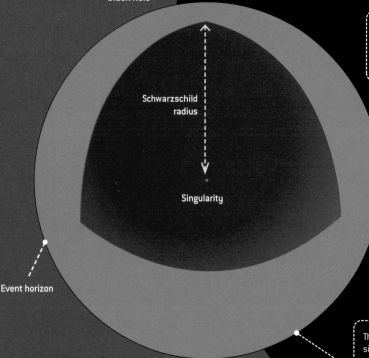

Schwarzschild radius

Singularity

Event horizon

Once an object has fallen inside the event horizon, space and time become interchangeable ... space-like paths become time-like — it is impossible to avoid the singularity at the centre.

These two descriptions of black holes are, in effect, simply mathematical solutions: mathematical surfaces and forms that predict almost unimaginable conditions. Black holes, or at least the effects of their intense gravitational fields on surrounding space, have now been directly observed.

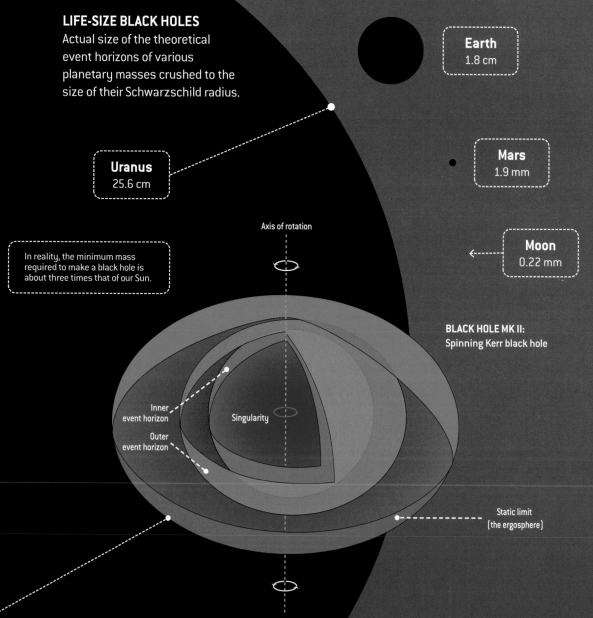

LIFE-SIZE BLACK HOLES
Actual size of the theoretical event horizons of various planetary masses crushed to the size of their Schwarzschild radius.

Earth
1.8 cm

Uranus
25.6 cm

Mars
1.9 mm

In reality, the minimum mass required to make a black hole is about three times that of our Sun.

Moon
0.22 mm

Axis of rotation

BLACK HOLE MK II:
Spinning Kerr black hole

Inner event horizon

Singularity

Outer event horizon

Static limit
(the ergosphere)

BLACK-HOLE WEIRDNESS

It wasn't until 1963 that mathematician Roy Kerr derived an Einsteinian solution for a rotating black hole: the maths was fiendishly difficult. As black holes are formed by the collapse of rotating stellar and galactic material, it's probably the case that most black holes do indeed have an angular momentum. Spinning Kerr black holes have some unusual properties:

- there is no point singularity at their centre, but rather a ring singularity – a spinning one-dimensional ring
- there are two event horizons, an inner and outer one, and an ellipsoid called the ergosphere, inside which spacetime rotates with the black hole faster than the speed of light
- after passing through the inner event horizon spacetime itself is reversed: gravity near the ring singularity becomes repulsive, actually pushing you away from the centre
- ring singularities can be linked through spacetime, so they can act as wormholes – travelling through a ring singularity might take you to another point in spacetime, such as another Universe.

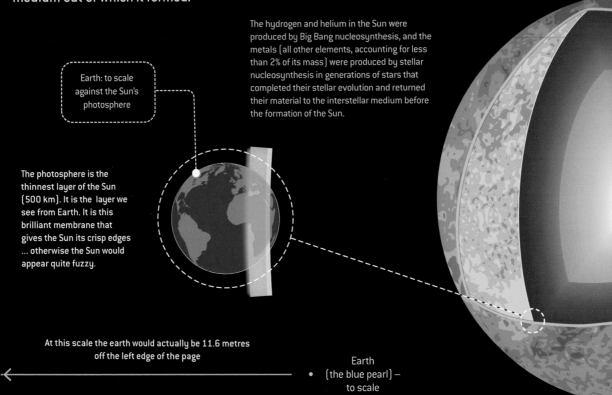

Our Sun is classed as a yellow dwarf star and is about 4.5 billion years old. It is now firmly in its middle age and will eventually expand to form a red giant in another 4 billion years. The Sun inherited its chemical composition from the interstellar medium out of which it formed.

The hydrogen and helium in the Sun were produced by Big Bang nucleosynthesis, and the metals (all other elements, accounting for less than 2% of its mass) were produced by stellar nucleosynthesis in generations of stars that completed their stellar evolution and returned their material to the interstellar medium before the formation of the Sun.

Earth: to scale against the Sun's photosphere

The photosphere is the thinnest layer of the Sun (500 km). It is the layer we see from Earth. It is this brilliant membrane that gives the Sun its crisp edges ... otherwise the Sun would appear quite fuzzy.

At this scale the earth would actually be 11.6 metres off the left edge of the page

Earth
(the blue pearl) –
to scale

The Sun
• 8.4 light minutes
from Earth
• 1 solar radius

It takes hundreds of thousands of years for photons to battle from core to the photosphere ... but neutrinos fly straight out from the centre at close to the speed of light.

Sirius
• 8.6 light years
from Earth
• 1.7 solar radius

Aldebaran
• 65 light years
from Earth
• 44.2 solar radius

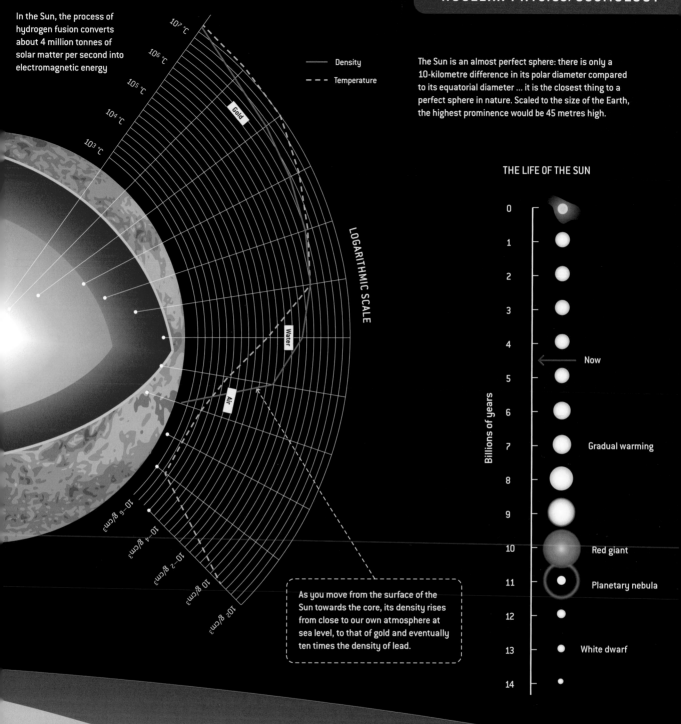

In the Sun, the process of hydrogen fusion converts about 4 million tonnes of solar matter per second into electromagnetic energy

Density

Temperature

The Sun is an almost perfect sphere: there is only a 10-kilometre difference in its polar diameter compared to its equatorial diameter ... it is the closest thing to a perfect sphere in nature. Scaled to the size of the Earth, the highest prominence would be 45 metres high.

LOGARITHMIC SCALE

10^7 °C
10^6 °C
10^5 °C
10^4 °C
10^3 °C

Gold

Water

Air

10^{-6} g/cm³
10^{-4} g/cm³
10^{-2} g/cm³
10 g/cm³
10^2 g/cm³

THE LIFE OF THE SUN

Billions of years

0
1
2
3
4
5 — Now
6
7 — Gradual warming
8
9
10 — Red giant
11 — Planetary nebula
12
13 — White dwarf
14

As you move from the surface of the Sun towards the core, its density rises from close to our own atmosphere at sea level, to that of gold and eventually ten times the density of lead.

The Pistol Star
• 25,000 light years from Earth
• 306 solar radius

Betelgeuse
• 640 light years from Earth
• 1,075 solar radius

3.14159265358979323846 26
51058209749445923078164
170679821480865132823066
35940812848111745028410 2
95493038196442881097566
31652712019091456485669 2
36072602491412737245870 0
282925409171536436789 25
21384146951941511609433 0
73819326117931051185480
2724891227938183011949 1
213949463952247371907 02
93176752384674818467669 4
27785771342757789609173 0
301465495858537105079 2279
219608640344181598136 29
9999837297804995105973 1
34690830264252230825 33
3137838752886587533208 3
3490428755468731159562 8
778053217122680661300192

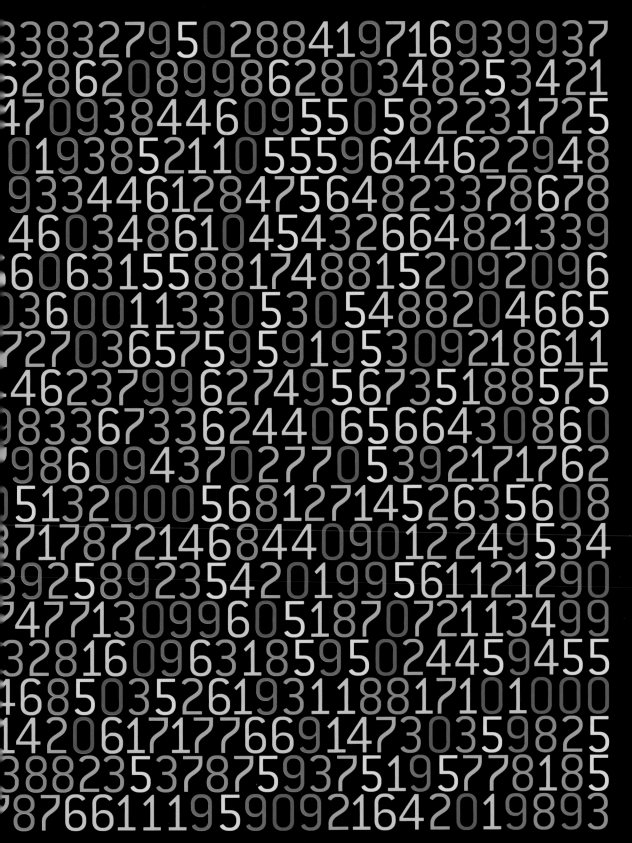

8832795028841971693993762862089986280348253421470938446095505822317250193852110555964462294893344612847564823378678460348610454326648213396063155881748815209609603600113053054882046657270365759591953092186114623799627495673518857508336733624406566430860986094370277053921717620513200056812714526356083717872146844090122495343925892354201995611212907477130996051870721134993281609631859502445945546850352619311881710100014206171776691473035982538822535787593751957781857876611195909216420198939

EARTH

Earth

About 4,600,000,000 years ago a small part of a local, giant molecular cloud began to collapse. Like an ice-dancer pulling in their arms to spin faster, as gravity pulled in the collapsing mass, tighter and tighter it began to spin with increasing speed. Most of the mass began collecting in the centre, rotation flattening out the remainder into a disc out of which the planets, moons, asteroids, and other small Solar System bodies began to form. Eventually, under increasing pressure and temperature the dense, whirling ball of hydrogen at the centre of the nascent Solar System spluttered into life – the Sun's furnace lit up. Over millions of years the dusty, gaseous disc began clumping into ever-larger planetesimals.

Though still composed mostly of hydrogen and helium, the stardust and gas forming this system contained a rich mix of heavier atoms and molecules: nitrogen, iron, nickel, phosphorus, water, methane ... forged in earlier supernovae and further melded by intense interstellar radiation over the many eons. Those heavier elements began to form solid metallic and silicate cores in the infant planets. The Solar System was still chaotic, hot, crowded: huge collisions and re-orientations were frequent. Local swirls and eddies focused the collapse. As the Sun's rays heated the inner Solar System, more volatile gases and liquid compounds were boiled away and blasted out towards the edge of the system. The inner forming planets were cooked to rocky lumps of relatively small size. Further out, beyond the frost line, huge balls of gas and dust accreted, liquified, solidified under pressure: the gas giants ... Jupiter, Saturn, Neptune.

The nascent planets grew, sweeping the space around them as they adopted their orbits around the ever-brightening Sun. It was still a time of turmoil. Our Earth, still glowing with heat, struck another planetary body, the huge collision dashing vast quantities of molten matter into the space around it. Slowly falling into orbit, the debris coalesced. Our Moon, the largest proportionately in the Solar System, looked down upon Earth.

As the planet cooled, heavy, pure molten metals – iron and nickel – sunk to form the Earth's core. Lighter rocks began to solidify on the surface, forming a crust over which liquid water began to condense, Earth's gravity holding a thin, hot, dense envelope of gases around itself. The crust started to break into plates that moved and jostled, sliding under each other in places, pulling apart in others: thus the slow turning gyres that power the Earth's magnetic fields and build our mountain ranges and deep river valleys began. As more and more water collected, the planet was cooling – awash with seas. The chemistry of the planet began to change. Water's unique properties, its density, stability and quality as a 'universal solvent', opened up new realms of aqueous chemistry. Rain eroded the newly thrown-up mountains, rivers carried water rich in mineral down to the ever-growing seas. The atmosphere cooled and became lighter as carbon dioxide was sequestered into the oceans and buried deep within the Earth – the complex system of weather evolved, cycling and recycling liquid water through the air, the seas and across the lands.

Where tectonic plates moved apart, heated gases and minerals escaped into the deep-sea waters: ever more complex chemical compounds, organic compounds and silicates evolved. Bubbling through chalky crevices, reacting, catalysing, creating a rich alchemist's brew of dissolved chemicals ...

And it was in this environment, deep in the bathysmal gloom, some six hundred million years after the birth of the Earth, that chemistry changed for ever. The Universe as we know it changed. Chemicals became ever more complex ... then they began to copy themselves, and code complexity, growing yet greater complexity. Life began.

Planet formation

Our Solar System began to form 4.6 billion years ago with the gravitational collapse of a small part of a giant molecular cloud.

The Sun contains 99.86% of the Solar System's mass

A gravitationally unstable area of interstellar gas and dust coalesces into a smaller denser clump, which then begins to rotate and collapse – the rotation causing it to flatten into a disc

Rocks and metals condense throughout the disc; water, methane, ammonia and other ices only condense in the outer parts

If the disc is massive enough, the runaway accretions begin, resulting in the rapid – 100,000 to 300,000 years – formation of Moon- to Mars-sized planetary embryos

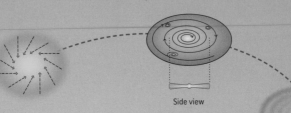

Side view

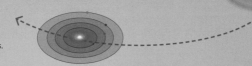

Near the burgeoning star, the planetary embryos go through a stage of violent mergers, producing a few terrestrial planets. The last stage takes approximately 100 million to a billion years

Mass of the Sun
1,989,000,000,000,000,000,000,000,000,000 kg

The surface area of this line represents the proportionate mass of the main planets, relative to the area of the Sun as depicted here

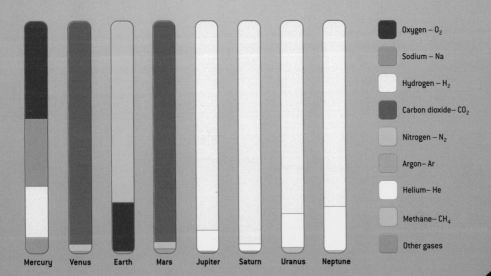

Legend:

- Oxygen – O_2
- Sodium – Na
- Hydrogen – H_2
- Carbon dioxide – CO_2
- Nitrogen – N_2
- Argon – Ar
- Helium – He
- Methane – CH_4
- Other gases

Planets: Mercury · Venus · Earth · Mars · Jupiter · Saturn · Uranus · Neptune

COMPOSITION OF THE PLANETARY ATMOSPHERES

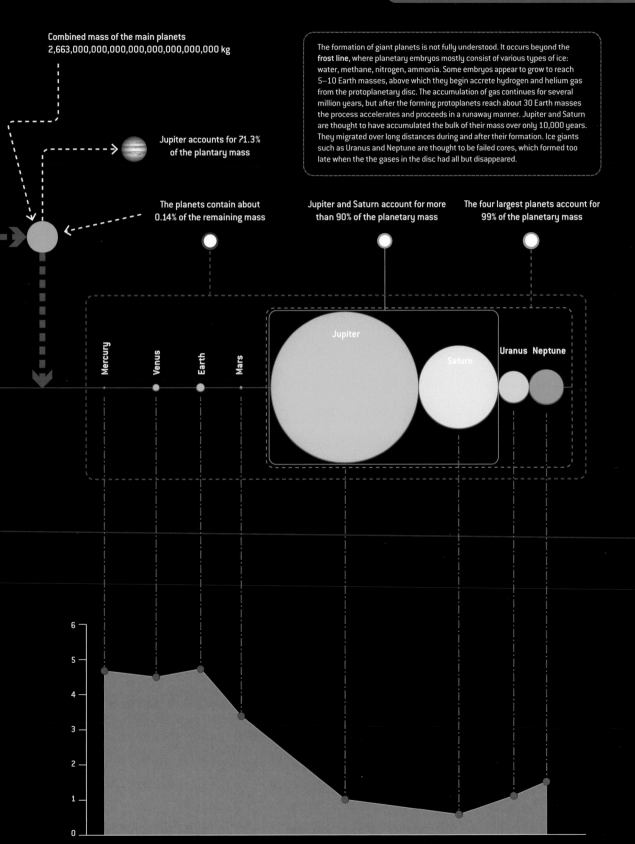

Combined mass of the main planets
2,663,000,000,000,000,000,000,000,000 kg

Jupiter accounts for 71.3%
of the plantary mass

The formation of giant planets is not fully understood. It occurs beyond the **frost line**, where planetary embryos mostly consist of various types of ice: water, methane, nitrogen, ammonia. Some embryos appear to grow to reach 5–10 Earth masses, above which they begin accrete hydrogen and helium gas from the protoplanetary disc. The accumulation of gas continues for several million years, but after the forming protoplanets reach about 30 Earth masses the process accelerates and proceeds in a runaway manner. Jupiter and Saturn are thought to have accumulated the bulk of their mass over only 10,000 years. They migrated over long distances during and after their formation. Ice giants such as Uranus and Neptune are thought to be failed cores, which formed too late when the the gases in the disc had all but disappeared.

The planets contain about
0.14% of the remaining mass

Jupiter and Saturn account for more
than 90% of the planetary mass

The four largest planets account for
99% of the planetary mass

Mercury

Venus

Earth

Mars

Jupiter

Saturn

Uranus Neptune

6

5

4

3

2

1

0

DENSITY OF THE PLANETS (RELATIVE TO JUPITER = 1)

The Solar System

A planet is defined as a celestial body in orbit around the Sun. It assumes a nearly round shape because it has sufficient mass for its self-gravity to overcome rigid body forces, and it has cleared the neighbourhood around its orbit of other solar bodies.

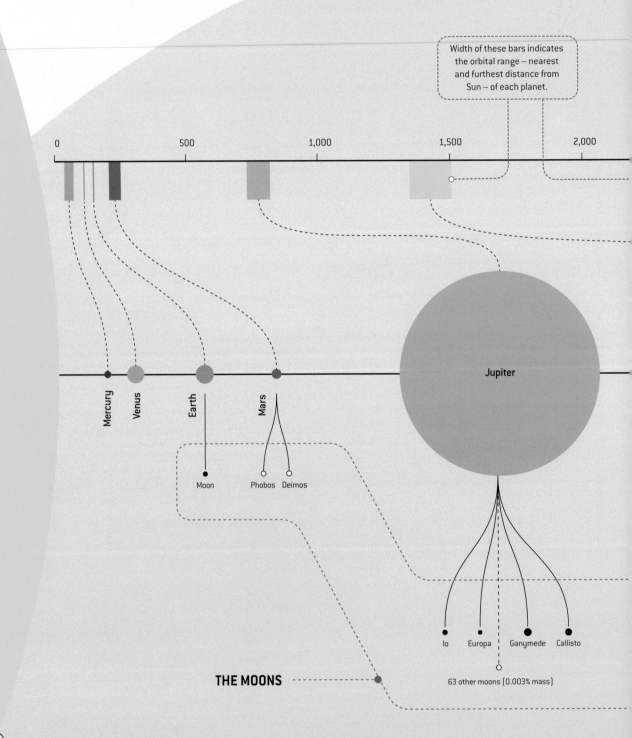

Width of these bars indicates the orbital range – nearest and furthest distance from Sun – of each planet.

0   500   1,000   1,500   2,000

Mercury

Venus

Earth

Mars

Jupiter

Moon

Phobos   Deimos

Io   Europa   Ganymede   Callisto

63 other moons (0.003% mass)

THE MOONS

There are 176 known natural moons orbiting planets in our Solar System. There are 168 moons orbiting the main planets — **Mercury, Venus, Earth, Mars, Jupiter, Saturn, Uranus, and Neptune** — while 8 moons orbit the dwarf planets — **Ceres, Pluto, Haumea, Makemak and Eris.**

SIZE OF SUN TO SCALE

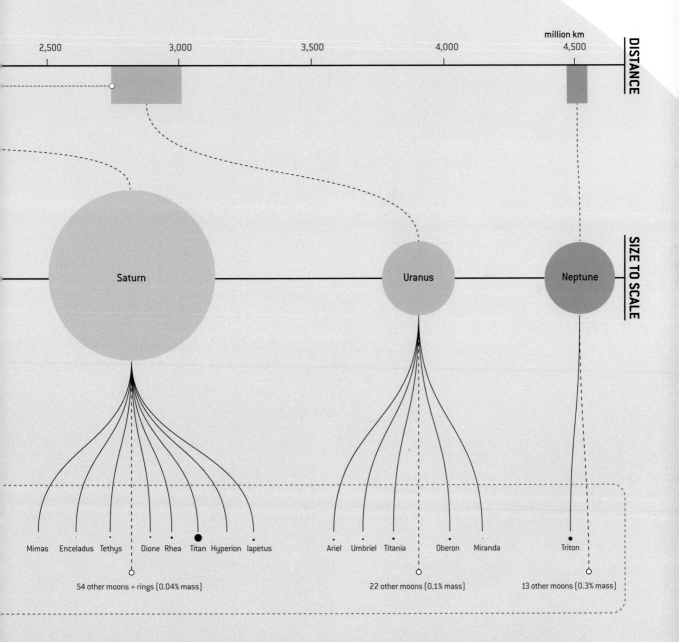

million km

DISTANCE

| 2,500 | 3,000 | 3,500 | 4,000 | 4,500 |

SIZE TO SCALE

Saturn

Uranus

Neptune

Mimas Enceladus Tethys Dione Rhea Titan Hyperion Iapetus

54 other moons + rings (0.04% mass)

Ariel Umbriel Titania Oberon Miranda

22 other moons (0.1% mass)

Triton

13 other moons (0.3% mass)

The Solar System: non-planetary bodies

The Solar System contains a vast array of orbiting bodies and extends thousands of times further out into interstellar space than the outer planets.

Eris is the most massive and the second-largest dwarf planet. It is the ninth most massive known body directly orbiting the Sun. Eris is 27% more massive than dwarf planet Pluto, though Pluto is slightly larger by volume – it was discovered in January 2005.

Possible orbit of 'Planet X'

Orbit of Pluto

Orbit of Eris

The Sun

Earth

Mars

Jupiter

Saturn

Uranus

Neptune

1

10

100

1,0

An Astronomical Unit is the distance from the Earth to the sun: it is used as a unit of distance

Heliosphere: charged particles stream out from the Sun at 1–3 million km/hour. Except for localised regions near obstacles such as planets or comets, this solar wind dominates the heliosphere

Heliopause: as the solar wind begins to interact with the interstellar medium, its velocity slows … eventually to a halt at the heliopause, where the interstellar medium and solar wind pressures balance

LOGARITHMIC PLAN OF THE SOLAR SYSTEM AND SURROUNDING SPACE

COMPARATIVE SCALE AND CLASSIFICATION

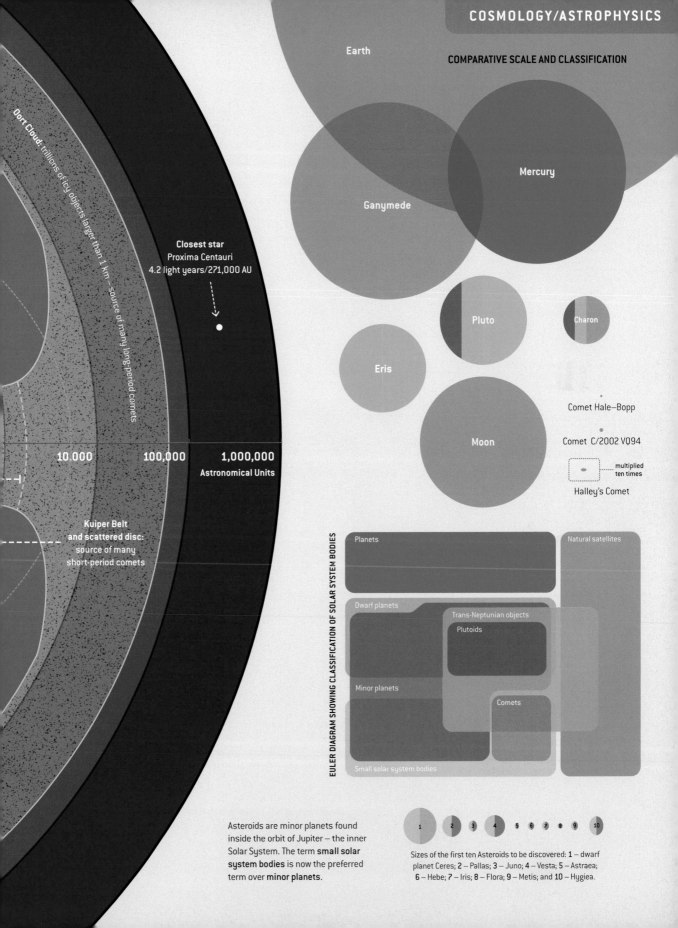

Earth

Mercury

Ganymede

Oort Cloud: trillions of icy objects larger than 1 km – source of many long-period comets

Closest star
Pluto
Proxima Centauri
4.2 light years/271,000 AU

Charon

Eris

Comet Hale–Bopp

Comet C/2002 VQ94

multiplied
ten times

Moon

Halley's Comet

10.000   100,000   1,000,000
Astronomical Units

**Kuiper Belt
and scattered disc:**
source of many
short-period comets

EULER DIAGRAM SHOWING CLASSIFICATION OF SOLAR SYSTEM BODIES

Planets

Natural satellites

Dwarf planets

Trans-Neptunian objects

Plutoids

Minor planets

Comets

Small solar system bodies

Asteroids are minor planets found
inside the orbit of Jupiter – the inner
Solar System. The term **small solar
system bodies** is now the preferred
term over **minor planets**.

1 2 3 4 5 6 7 8 9 10

Sizes of the first ten Asteroids to be discovered: 1 – dwarf
planet Ceres; 2 – Pallas; 3 – Juno; 4 – Vesta; 5 – Astraea;
6 – Hebe; 7 – Iris; 8 – Flora; 9 – Metis; and 10 – Hygiea.

The Moon and other moons

The Earth's Moon is the most massive in the Solar System in comparison to its planet. Amongst the moons of the main planets, its nearest rival is Saturn's Titan, which is around one hundred times smaller, proportionately. There are 173 known moons orbiting six of the eight planets.

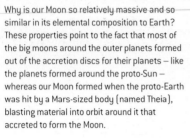

Why is our Moon so relatively massive and so similar in its elemental composition to Earth? These properties point to the fact that most of the big moons around the outer planets formed out of the accretion discs for their planets – like the planets formed around the proto-Sun – whereas our Moon formed when the proto-Earth was hit by a Mars-sized body (named Theia), blasting material into orbit around it that accreted to form the Moon.

The US Apollo program achieved six manned lunar landings between 1969 and 1972. These missions returned over 380 kg of lunar rocks, which have been analysed to provide a geological understanding of the Moon's origin.

As the Moon pulls the Earth's tides around the globe, frictional drag is very gradually slowing the speed of the Moon's orbit. As a result, the Moon is moving away from the Earth at rate of 3.8 cm per year.

CHEMICAL COMPOSITION OF THE EARTH AND THE MOON

- Oxygen
- Silicon
- Aluminium
- Iron
- Calcium
- Sodium
- Potassium
- Magnesium
- Titanium
- Other elements

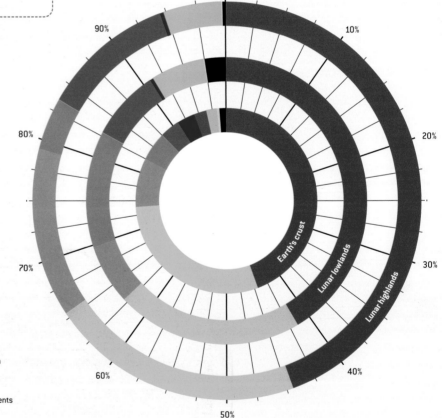

THE EARTH, THE MOON, AND THE SPACE BETWEEN ... TO SCALE

ACTUAL SIZES OF THE LARGEST MOONS

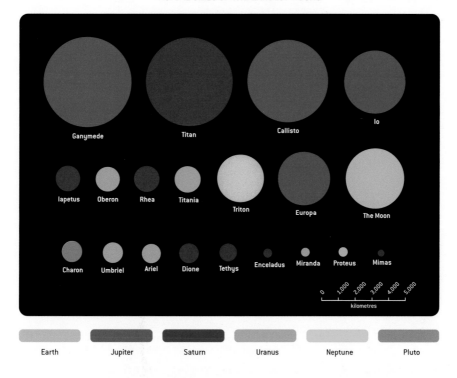

Earth   Jupiter   Saturn   Uranus   Neptune   Pluto

MASS OF MOONS AS PERCENTAGE OF PLANET MASS (LOGARITHMIC SCALE)

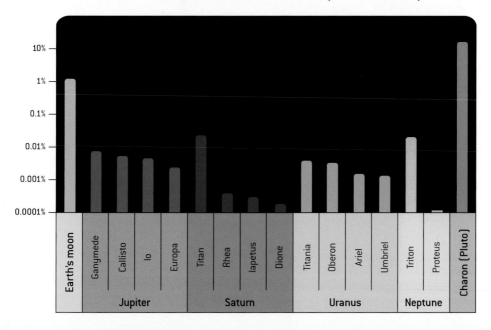

The Earth

About half the internal heat of the Earth is primordial heat, and half is generated by the continuing decay of radioactive metals such as uranium, thorium and potassium. The additional heat this decay provides is vital in powering life-nurturing and life-sustaining processes such as the Earth's magnetic field, plate tectonics and volcanism.

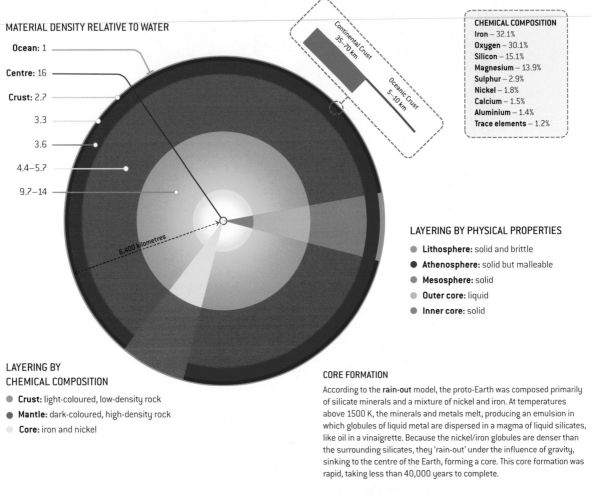

MATERIAL DENSITY RELATIVE TO WATER

Ocean: 1
Centre: 16
Crust: 2.7
3.3
3.6
4.4–5.7
9.7–14

6,400 kilometres

Continental Crust
35–70 km

Oceanic Crust
5–10 km

CHEMICAL COMPOSITION
Iron – 32.1%
Oxygen – 30.1%
Silicon – 15.1%
Magnesium – 13.9%
Sulphur – 2.9%
Nickel – 1.8%
Calcium – 1.5%
Aluminium – 1.4%
Trace elements – 1.2%

LAYERING BY PHYSICAL PROPERTIES
- **Lithosphere:** solid and brittle
- **Athenosphere:** solid but malleable
- **Mesosphere:** solid
- **Outer core:** liquid
- **Inner core:** solid

LAYERING BY
CHEMICAL COMPOSITION
- **Crust:** light-coloured, low-density rock
- **Mantle:** dark-coloured, high-density rock
- **Core:** iron and nickel

CORE FORMATION

According to the **rain-out** model, the proto-Earth was composed primarily of silicate minerals and a mixture of nickel and iron. At temperatures above 1500 K, the minerals and metals melt, producing an emulsion in which globules of liquid metal are dispersed in a magma of liquid silicates, like oil in a vinaigrette. Because the nickel/iron globules are denser than the surrounding silicates, they 'rain-out' under the influence of gravity, sinking to the centre of the Earth, forming a core. This core formation was rapid, taking less than 40,000 years to complete.

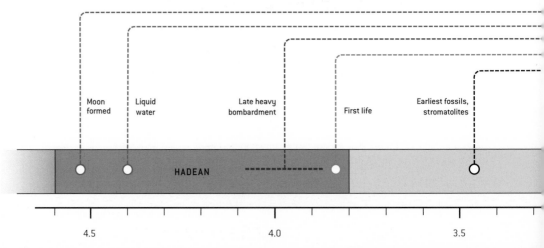

Moon
formed

Liquid
water

Late heavy
bombardment

First life

Earliest fossils,
stromatolites

HADEAN

4.5   4.0   3.5

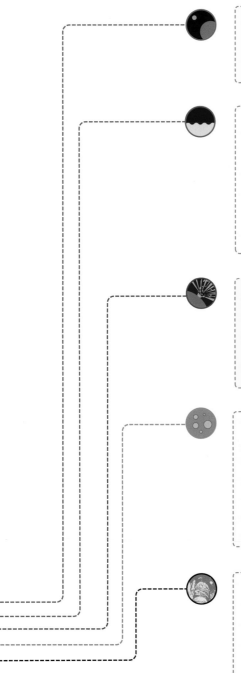

MOON FORMED

The Moon formed out of debris resulting from an indirect collision between Earth and an astronomical body the size of Mars, approximately 4.5 billion years ago, in the Hadean eon — about 20 to 100 million years after the Solar System coalesced.

LIQUID WATER

While water is common throughout the Universe — the product of Big Bang hydrogen and oxygen forged in the heart of dying stars — and would have been a notable constituent of the nascent Solar System, the origin of water on Earth remains uncertain. After formation, the Earth had no atmosphere and any liquid primordial water would have boiled out into space. It is thought that the bulk of water on Earth was delivered by water-rich protoplanets formed in the outer asteroid belt that plunged towards the Earth and continued to collide with the planet for hundreds of millions of years after its formation. Alternative theories, based on the compositon of certain meteorites and asteroids with high water content (up to 20%), suggest that water on Earth may have accreted at the same time as the rock.

LATE HEAVY BOMBARDMENT

Between 4.1 and 3.8 billion years ago, there was a period when a disproportionately large number of asteroids collided with the planets in the inner Solar System — Mercury, Venus, Earth (and its Moon) and Mars. This occurred after the planets had formed and accreted most of their mass. It is thought that these events may have given early life a boost by 'pruning' the Tree of Life and inducing hydrothermal systems, forcing life into sub-surface and submarine havens that provided a crucible for the emergence of more developed cells.

FIRST LIFE

The evidence of first life on Earth emerges in the record about 3.7 billion years ago — initially with single-celled prokaryotic cells, such as bacteria. It took more than another billion years before true multicellular life evolved, and it's only in the last 570 million years, that the early ancestors of familiar modern life forms began to emerge, the first being arthropods (insects/crustacea). The earliest evidence of life currently is biogenic graphite found in 3.7-billion-year-old sedimentary rocks discovered in western Greenland. Carbon derived from living sources tends to be depleted in the heavier carbon-13, due to preferential take-up of the lighter carbon-12

FIRST FOSSILS

Cyanobacteria, previously called blue-green algae, are an archaic, large form of bacteria, and can secrete a thick cell wall. More importantly, they may form large dome-shaped, layered structures, called **stromatolites**. These structures form from a mat of cyanobacteria in an aquatic environment, trapping sediment and sometimes secreting calcium carbonate. When sectioned very thinly, fossil stromatolites may be found to contain exquisitely preserved fossil cyanobacteria and algae. The oldest cyanobacteria-like fossils known are nearly 3.5 billion years old.

ARCHEAN		PRECAMBRIAN

3.0   2.5   Billions of years before present

The lithosphere

The lithosphere is the rigid outer layer of the Earth, consisting of the crust — continental and oceanic — and an underlying rigid region of the mantle. The lithosphere rides on top of a viscous part of the mantle called the asthenosphere.

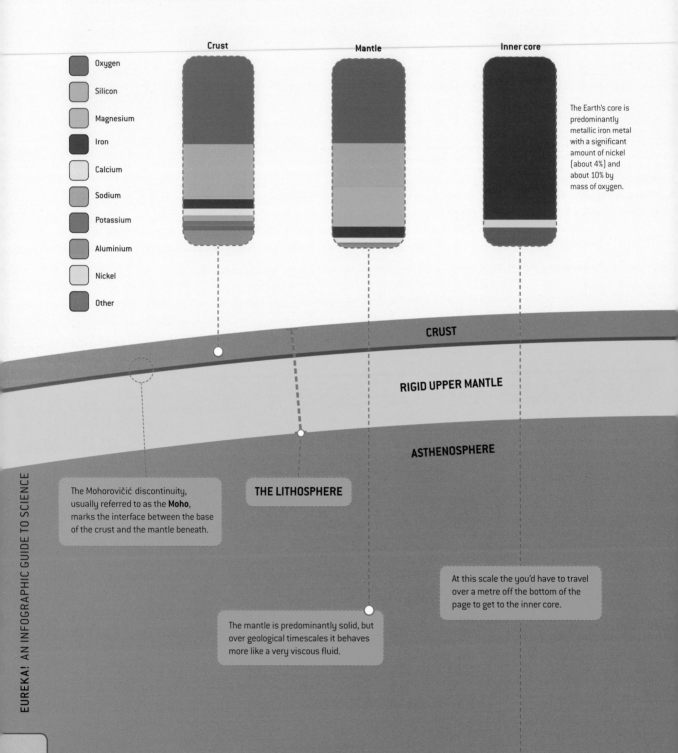

Crust

Mantle

Inner core

Oxygen
Silicon
Magnesium
Iron
Calcium
Sodium
Potassium
Aluminium
Nickel
Other

The Earth's core is predominantly metallic iron metal with a significant amount of nickel (about 4%) and about 10% by mass of oxygen.

CRUST

RIGID UPPER MANTLE

ASTHENOSPHERE

THE LITHOSPHERE

The Mohorovičić discontinuity, usually referred to as the **Moho**, marks the interface between the base of the crust and the mantle beneath.

The mantle is predominantly solid, but over geological timescales it behaves more like a very viscous fluid.

At this scale the you'd have to travel over a metre off the bottom of the page to get to the inner core.

EARTH'S INTERNAL CONVECTION AND TECTONICS

Mantle convection is the slow creeping motion of Earth's silicate mantle caused by convection currents carrying heat from the interior of the Earth to the surface. The overlying lithosphere is divided into a number of plates that are continuously being created and consumed at their opposite plate boundaries. Large convection currents in the asthenosphere transfer heat to the surface, where plumes of magma break apart the plates at the spreading centres, creating divergent plate boundaries. As the plates move away, they cool and get consumed at the ocean trenches/subduction zones. The crust is recycled back into the asthenosphere. Typical mantle convection speed is 20 mm per year near the crust but much slower near the core. A single shallow convection cycle might take around 50 million years, though deeper convection cyles can be closer to 200 million years.

At the heart of the Earth is a solid inner core two-thirds the size of the Moon and as hot as the surface of the Sun – the extreme pressure prevents it from becoming liquid. Surrounding this is the outer core where lower pressure means the iron here is fluid. Convection flows within this liquid iron generate electric currents, which in turn produce magnetic fields. This is the source of the life-preserving magnetic field that surrounds Earth.

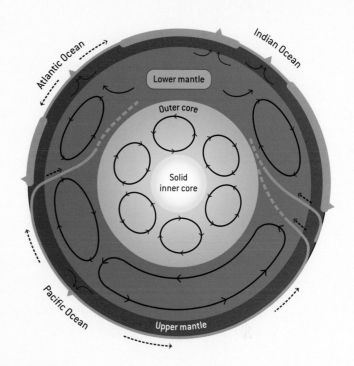

DEEPEST MAN-MADE HOLES
We've barely scratched the surface

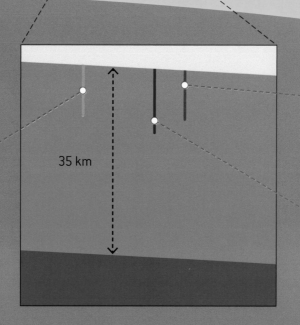

Longer boreholes have been drilled for oil extraction, but often there is a considerable sideways element to the drilling. These examples are attempts to drill vertically down into the crust.

35 km

Bertha Rogers Hole
An oil-exploratory hole drilled in Washita County, Oklahoma. The well encountered enormous pressure – almost 2,000 times atmospheric pressure. Drilling ceased when the drilling hit a molten sulphur deposit, which destroyed the drill bit.
Depth: 9.6 kilometres
Temperature: 246 ˚C

The KTB superdeep borehole
German Continental Deep Drilling Program in northern Bavaria.
Depth: 9.1 kilometres
Temperature: 260 ˚C

The Kola Superdeep Borehole
Soviet scientific drilling project on the Kola Peninsula, in the Russian Arctic.
Depth: 12.3 kilometres
Temperature: 180 ˚C

Plate tectonics

Tectonic activity, which describes the interaction of the huge slabs of lithosphere called tectonic plates, is responsible for some of Earth's most dramatic geological events: orogeny, earthquakes and volcanic activity. Orogeny is the process of mountain building.

Most tectonic activity takes place at the edges of these plates, where they collide, tear apart or slide against each other. The movement of tectonic plates is driven by the internal heat of the Earth and convection currents in the plastic rocks of the upper mantle.

The East Pacific Rise – so-called because it has gentler slopes than the Mid-Atlantic Ridge – spreads at rates of 6 to 16 cm per year. In the past the rate was even higher, at 20 cm per year. Due to the fast spreading rates, there is no rift valley, just a smooth volcanic summit with a crack along the crest that is much smaller than the Atlantic rift valley.

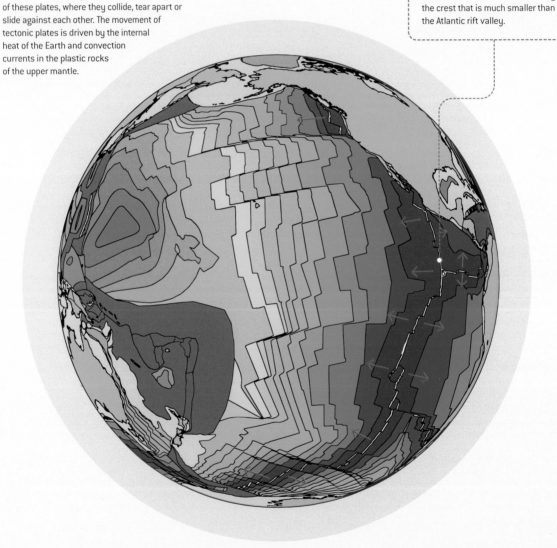

AGE OF SEABED

| 180 | 170 | 160 | 150 | 140 | 130 | 120 | 110 | 100 |

MILLIONS OF YEARS BEFORE THE PRESENT

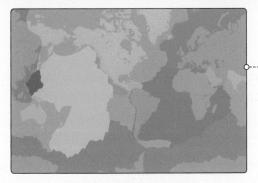

THE EARTH'S TECTONIC PLATES

The lithosphere is divided into fifteen major tectonic plates: the North American, Caribbean, South American, Scotia, Antarctic, Eurasian, Arabian, African, Indian, Philippine, Australian, Pacific, Juan de Fuca, Cocos, and Nazca.

The Mid-Atlantic Ridge runs down the centre of the Atlantic Ocean and spreads apart at rates of 2 to 5 cm per year. The ridge has a deep rift valley running along its crest that is 1 to 3 km deep – it is about the depth and width of the Grand Canyon.

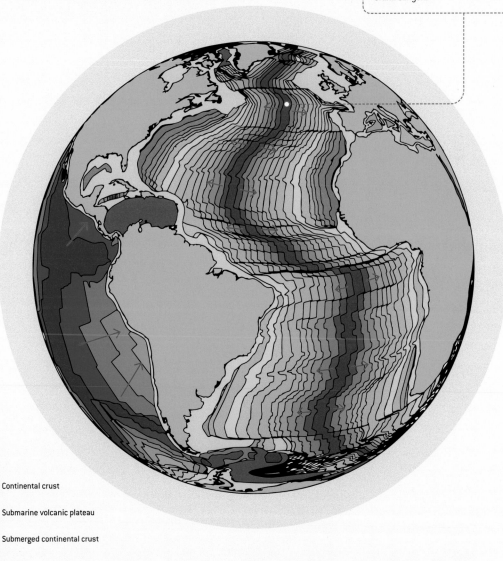

Continental crust

Submarine volcanic plateau

Submerged continental crust

PRESENT DAY

90   80   70   60   50   40   30   20   10

Earthquakes

As the Earth's tectonic plates move against each other – pulling apart in places, pushing along, against and under each other – they can lock, and elastic tension builds, storing huge amounts of potential energy. When the fault finally gives, the energy is released as a combination of radiated elastic strain seismic waves, frictional heating of the fault surface, and cracking of the crustal rock.

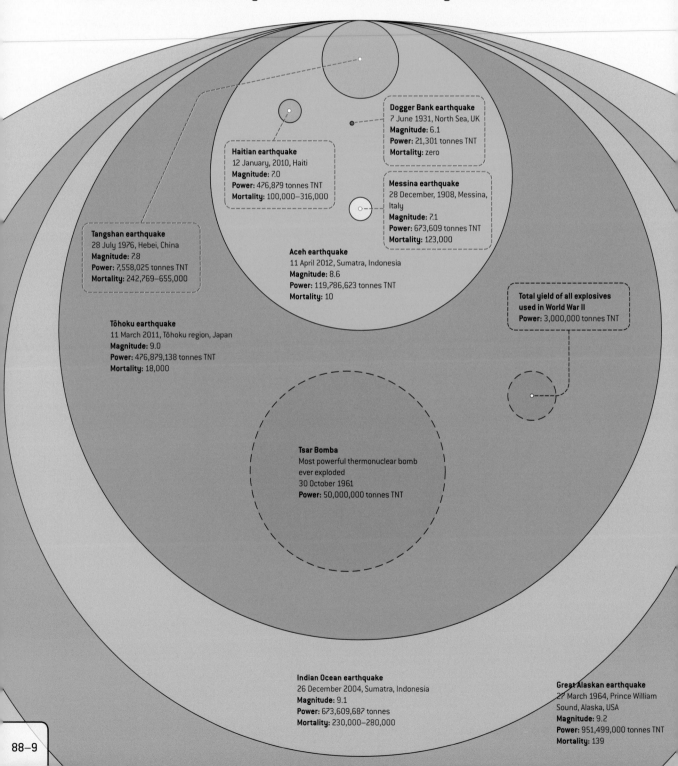

Dogger Bank earthquake
7 June 1931, North Sea, UK
Magnitude: 6.1
Power: 21,301 tonnes TNT
Mortality: zero

Haitian earthquake
12 January, 2010, Haiti
Magnitude: 7.0
Power: 476,879 tonnes TNT
Mortality: 100,000–316,000

Messina earthquake
28 December, 1908, Messina, Italy
Magnitude: 7.1
Power: 673,609 tonnes TNT
Mortality: 123,000

Tangshan earthquake
28 July 1976, Hebei, China
Magnitude: 7.8
Power: 7,558,025 tonnes TNT
Mortality: 242,769–655,000

Aceh earthquake
11 April 2012, Sumatra, Indonesia
Magnitude: 8.6
Power: 119,786,623 tonnes TNT
Mortality: 10

Total yield of all explosives used in World War II
Power: 3,000,000 tonnes TNT

Tōhoku earthquake
11 March 2011, Tōhoku region, Japan
Magnitude: 9.0
Power: 476,879,138 tonnes TNT
Mortality: 18,000

Tsar Bomba
Most powerful thermonuclear bomb ever exploded
30 October 1961
Power: 50,000,000 tonnes TNT

Indian Ocean earthquake
26 December 2004, Sumatra, Indonesia
Magnitude: 9.1
Power: 673,609,687 tonnes
Mortality: 230,000–280,000

Great Alaskan earthquake
27 March 1964, Prince William Sound, Alaska, USA
Magnitude: 9.2
Power: 951,499,000 tonnes TNT
Mortality: 139

Seismic waves are mechanical energy waves that transmit the energy released by violent tectonic events. The speed and range of these waves depend on the amount of energy released and the density and elasticity of the material through which the waves travel – ranging in speed between 7,200 and 28,000 kilometres per hour within the crust and 45,000 kilometres per hour in the deeper mantle.

TYPES OF SEISMIC WAVE

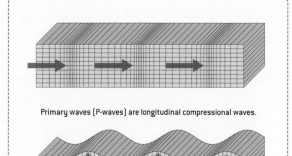

Primary waves (P-waves) are longitudinal compressional waves.

BODY WAVES:
travel through the interior of the Earth at great speed

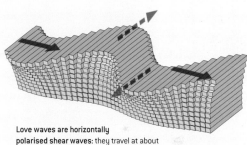

Secondary waves (S-waves) are transverse shear waves.

SURFACE WAVES:
diminish as they get further from the surface – they travel more slowly than seismic body waves

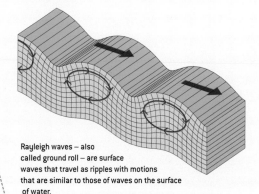

Love waves are horizontally polarised shear waves: they travel at about 60% of P-wave velocity and have the largest seismic amplitude.

Rayleigh waves – also called ground roll – are surface waves that travel as ripples with motions that are similar to those of waves on the surface of water.

Valdivia earthquake
22 May 1960, Valdivia, Chile
Magnitude: 9.5
Power: 2,681,688,466 tonnes TNT
Mortality: 2,230–6,000

It is estimated that around 500,000 detectable earthquakes occur globally each year. Of these, some 100,000 can be felt by humans.

The atmosphere surrounds us, it is literally the air that we breath — and yet this life-filled envelope seems a tenous skin when viewed at Earth scale.

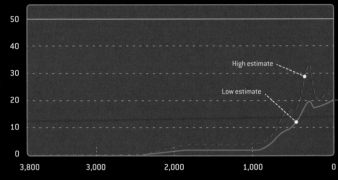

Percentage volume

50
40
30 — High estimate
20
10 — Low estimate
0

3,800   3,000   2,000   1,000   0

Millions of years before present

If the Earth was the size of a standard soccer ball:

- the troposphere would be 0.3 mm thick
- the stratosphere would reach up 1 mm
- the top of the mesosphere is at 1.4 mm

... and the International Space Station would be orbiting 7 mm off the surface of the ball.

Thermosphere

Mesosphere

Stratosphere

Troposphere

Kármán Line — beginning of space

99.99997% is below 100 km.

50% is below 5.6 km.

90% is below 16 km.

Proportion of atmosphere

Curvature of Earth is exaggerated ×10 — at this scale the Earth would actually be 13 metres across ... the length of a shipping container

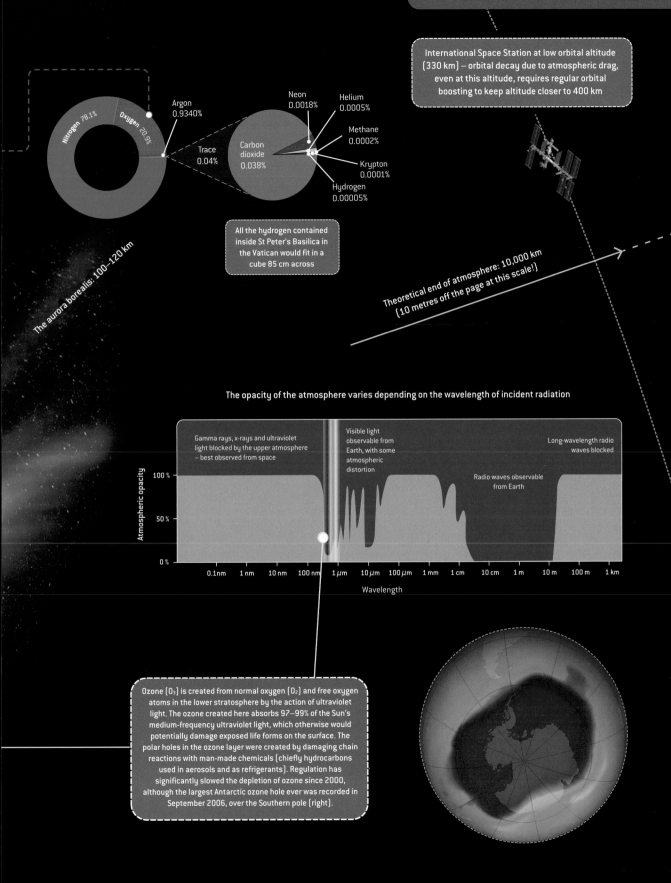

Nitrogen 78.1%

Oxygen 20.9%

Argon 0.9340%

Trace 0.04%

Neon 0.0018%

Helium 0.0005%

Methane 0.0002%

Krypton 0.0001%

Hydrogen 0.00005%

Carbon dioxide 0.038%

International Space Station at low orbital altitude (330 km) – orbital decay due to atmospheric drag, even at this altitude, requires regular orbital boosting to keep altitude closer to 400 km

All the hydrogen contained inside St Peter's Basilica in the Vatican would fit in a cube 85 cm across

Theoretical end of atmosphere: 10,000 km (10 metres off the page at this scale!)

The aurora borealis: 100–120 km

The opacity of the atmosphere varies depending on the wavelength of incident radiation

Gamma rays, x-rays and ultraviolet light blocked by the upper atmosphere – best observed from space

Visible light observable from Earth, with some atmospheric distortion

Long-wavelength radio waves blocked

Radio waves observable from Earth

Atmospheric opacity

100 %

50 %

0 %

0.1nm 1 nm 10 nm 100 nm 1 μm 10 μm 100 μm 1 mm 1 cm 10 cm 1 m 10 m 100 m 1 km

Wavelength

Ozone (O_3) is created from normal oxygen (O_2) and free oxygen atoms in the lower stratosphere by the action of ultraviolet light. The ozone created here absorbs 97–99% of the Sun's medium-frequency ultraviolet light, which otherwise would potentially damage exposed life forms on the surface. The polar holes in the ozone layer were created by damaging chain reactions with man-made chemicals (chiefly hydrocarbons used in aerosols and as refrigerants). Regulation has significantly slowed the depletion of ozone since 2000, although the largest Antarctic ozone hole ever was recorded in September 2006, over the Southern pole (right).

Climate zones: atmospheric circulation

While smaller-scale weather systems – mid-latitude depressions, troughs of high pressure and weather fronts – occur randomly, the large-scale structure of the atmosphere and its circulation remains fairly constant.

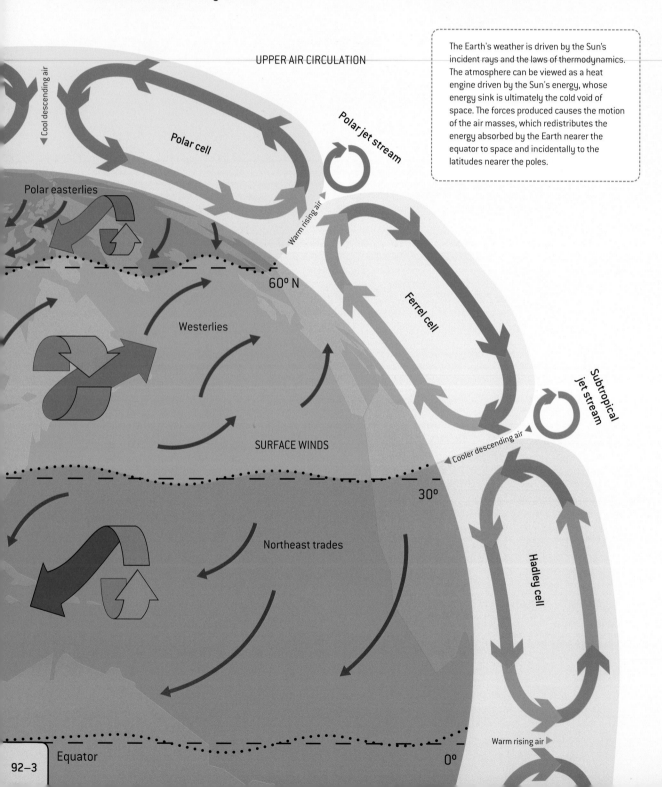

UPPER AIR CIRCULATION

Cool descending air

Polar cell

Polar jet stream

The Earth's weather is driven by the Sun's incident rays and the laws of thermodynamics. The atmosphere can be viewed as a heat engine driven by the Sun's energy, whose energy sink is ultimately the cold void of space. The forces produced causes the motion of the air masses, which redistributes the energy absorbed by the Earth nearer the equator to space and incidentally to the latitudes nearer the poles.

Polar easterlies

Warm rising air

Ferrel cell

60º N

Westerlies

Subtropical jet stream

Cooler descending air

SURFACE WINDS

30º

Northeast trades

Hadley cell

Warm rising air

Equator

0º

Averaged January wind speed and direction

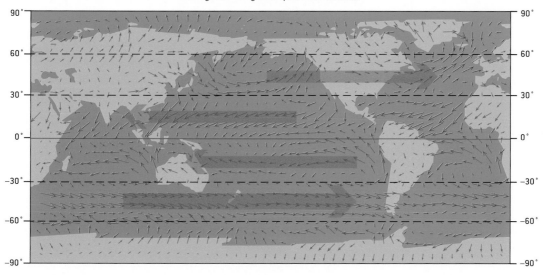

SURFACE WINDS

The cellular circulation of the Earth's atmosphere drives surface winds in latitudinal bands: westerly in the higher latitudes, easterly in the subtropical latitudes.

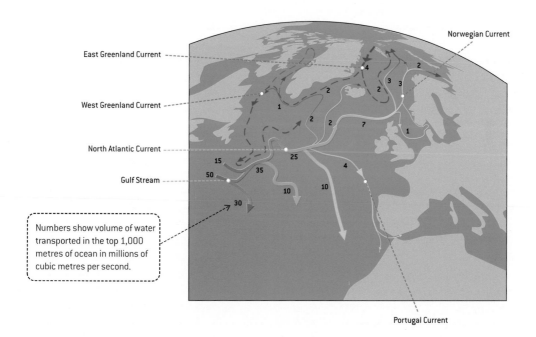

Numbers show volume of water transported in the top 1,000 metres of ocean in millions of cubic metres per second.

OCEAN CURRENTS

The oceans are the other main driver of weather systems. The prevailing winds drive huge movements of water, part of the same heat engine distributing and dissipating the Sun's energy. In the North Atlantic, currents draw waters up from the American Gulf and east coast up and across the ocean, carrying nutrients and warmth to higher latitudes and greatly affecting the climate in northwest Europe. Blanc-Sablon, on Canada's east coast, has January and July average temperatures of −10°C and 12°C; London, at the same latitude, has averages in the same months of 7°C and 19°C.

Meteorology

An air mass is a body of air within which variations in temperature and humidity are relatively slight – the air making up the mass is quite distinctive in temperature and humidity. Air masses are separated from adjacent bodies of air by weather fronts.

SURFACE PRESSURE AND AIR MASS MOVEMENT

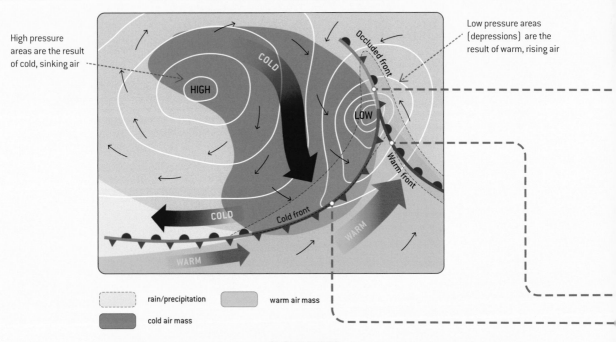

High pressure areas are the result of cold, sinking air

Low pressure areas (depressions) are the result of warm, rising air

HIGH

LOW

COLD

COLD

WARM

WARM

Occluded front

Warm front

Cold front

rain/precipitation

cold air mass

warm air mass

CLOUD TYPES

- **Cirrus** – a tuft or filament like hair
- **Cumulus** – heaped or piled
- **Stratus** – layered
- **Nimbus** – rain-bearing
- **Alto** – high up

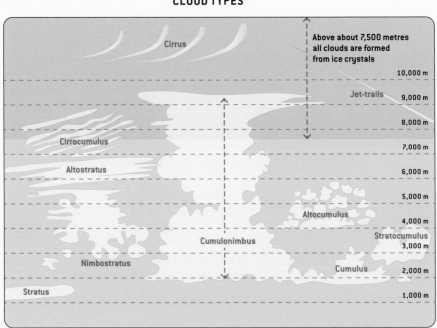

Above about 7,500 metres all clouds are formed from ice crystals

Cirrus

Jet-trails

10,000 m
9,000 m
8,000 m
7,000 m
6,000 m
5,000 m
4,000 m
3,000 m
2,000 m
1,000 m

Cirrocumulus

Altostratus

Altocumulus

Stratocumulus

Cumulonimbus

Cumulus

Nimbostratus

Stratus

CROSS SECTIONS OF WEATHER FRONTS

OCCLUDED FRONT

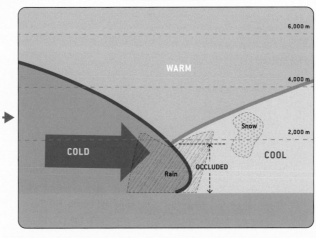

WARM FRONT

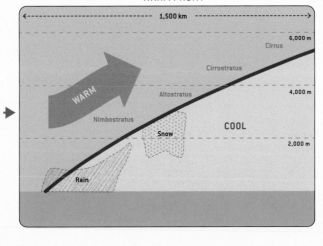

COLD FRONT

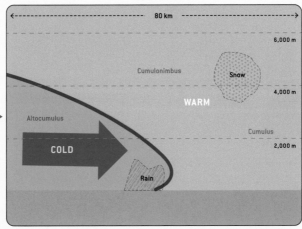

Climate extremes

Despite lethal extremes, the Earth's atmosphere is pretty temperate and relaxed by interplanetary standards.

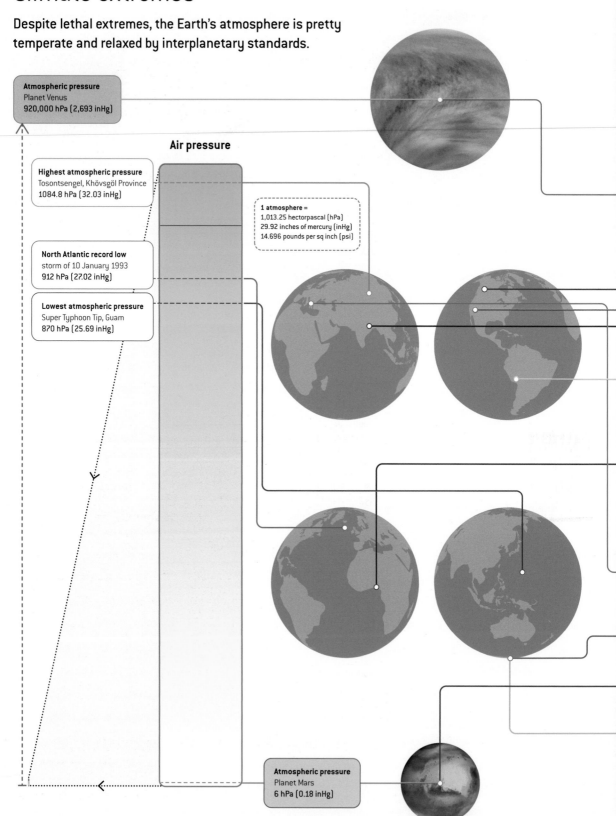

Atmospheric pressure
Planet Venus
920,000 hPa (2,693 inHg)

Air pressure

Highest atmospheric pressure
Tosontsengel, Khövsgöl Province
1084.8 hPa (32.03 inHg)

1 atmosphere =
1,013.25 hectorpascal (hPa)
29.92 inches of mercury (inHg)
14.696 pounds per sq inch (psi)

North Atlantic record low
storm of 10 January 1993
912 hPa (27.02 inHg)

Lowest atmospheric pressure
Super Typhoon Tip, Guam
870 hPa (25.69 inHg)

Atmospheric pressure
Planet Mars
6 hPa (0.18 inHg)

Weather generally refers to daily temperature and precipitation, whereas **climate** concerns the statistics of atmospheric conditions over longer periods of time. Weather is driven by the air pressure, temperature and moisture differences between areas of the planet. Surface temperatures generally range by plus or minus 40 °C. Surface temperature differences in turn cause pressure differences – warm air expands and lowers in density and resultant surface air pressure. The resulting horizontal pressure gradient moves the air from higher to lower pressure regions, creating wind, and the Earth's rotation then causes deflection of this air flow due to the Coriolis effect. The simple systems thus formed can then display emergent behaviour to produce more complex systems – the system is inherently chaotic, so small changes can grow to have large effects on the system as a whole.

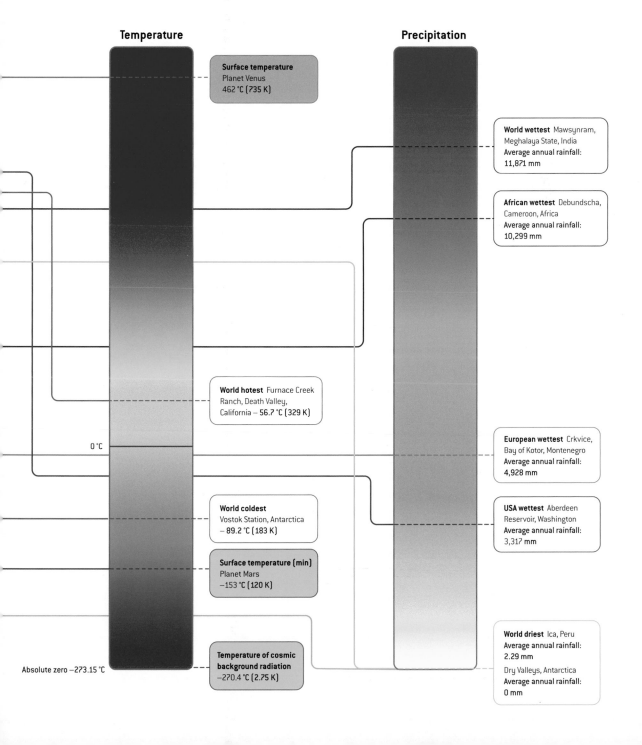

Temperature

Precipitation

Surface temperature
Planet Venus
462 °C (735 K)

World wettest Mawsynram, Meghalaya State, India
Average annual rainfall: 11,871 mm

African wettest Debundscha, Cameroon, Africa
Average annual rainfall: 10,299 mm

World hotest Furnace Creek Ranch, Death Valley, California – 56.7 °C (329 K)

0 °C

European wettest Crkvice, Bay of Kotor, Montenegro
Average annual rainfall: 4,928 mm

World coldest
Vostok Station, Antarctica
– 89.2 °C (183 K)

USA wettest Aberdeen Reservoir, Washington
Average annual rainfall: 3,317 mm

Surface temperature (min)
Planet Mars
−153 °C (120 K)

World driest Ica, Peru
Average annual rainfall: 2.29 mm

Dry Valleys, Antarctica
Average annual rainfall: 0 mm

Absolute zero −273.15 °C

Temperature of cosmic background radiation
−270.4 °C (2.75 K)

Solar activity and climate

The Sun doesn't burn with a constant flame, its output and brightness vary over time. There is a regular 11-year pulse of activity laid over more unpredictable, longer-term changes. The prevalence of sunspots was noticed by early astronomers and they have been counted meticulously since the mid-eighteenth century. They provide a good indicator for solar activity.

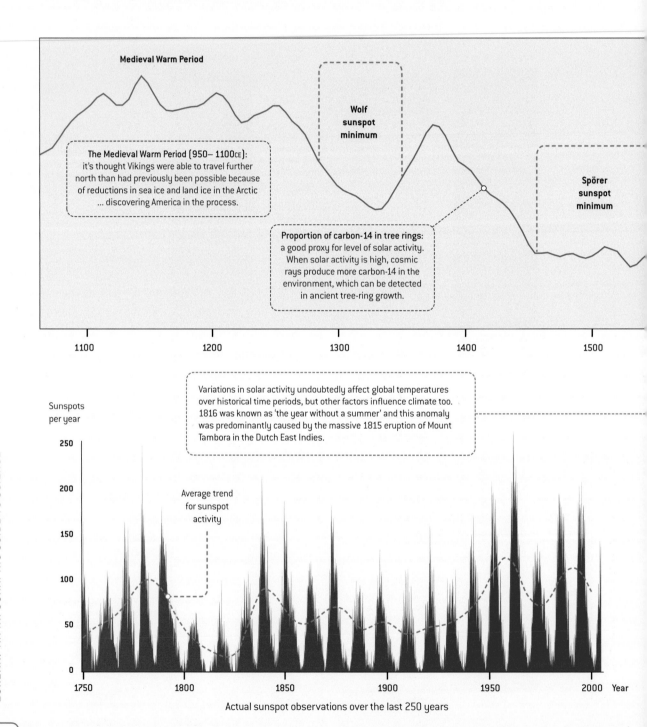

Medieval Warm Period

Wolf sunspot minimum

Spörer sunspot minimum

The Medieval Warm Period (950–1100CE): it's thought Vikings were able to travel further north than had previously been possible because of reductions in sea ice and land ice in the Arctic ... discovering America in the process.

Proportion of carbon-14 in tree rings: a good proxy for level of solar activity. When solar activity is high, cosmic rays produce more carbon-14 in the environment, which can be detected in ancient tree-ring growth.

1100   1200   1300   1400   1500

Variations in solar activity undoubtedly affect global temperatures over historical time periods, but other factors influence climate too. 1816 was known as 'the year without a summer' and this anomaly was predominantly caused by the massive 1815 eruption of Mount Tambora in the Dutch East Indies.

Sunspots per year

Average trend for sunspot activity

250

200

150

100

50

0

1750   1800   1850   1900   1950   2000   Year

Actual sunspot observations over the last 250 years

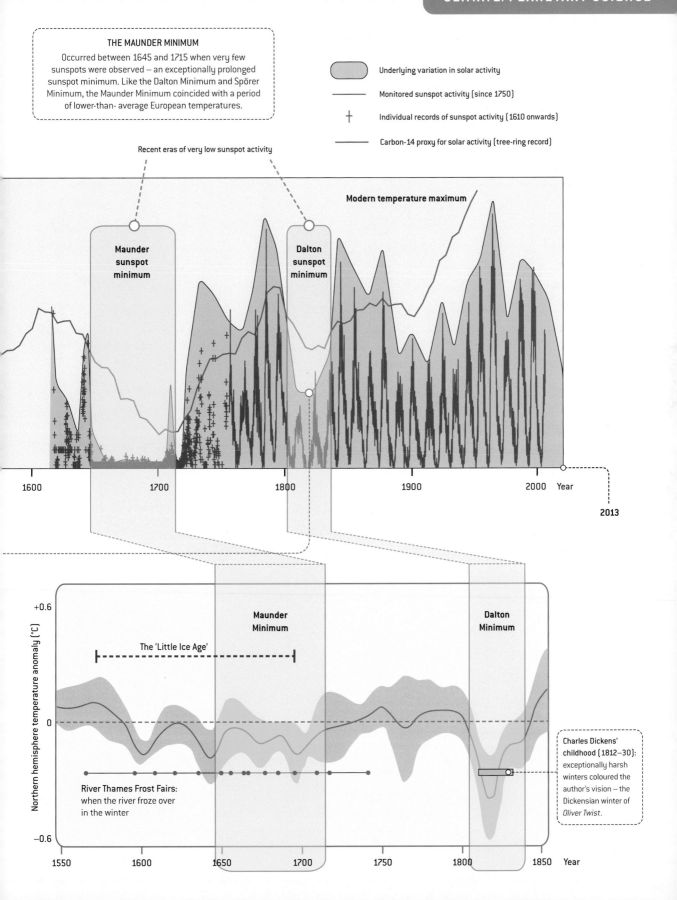

THE MAUNDER MINIMUM
Occurred between 1645 and 1715 when very few sunspots were observed – an exceptionally prolonged sunspot minimum. Like the Dalton Minimum and Spörer Minimum, the Maunder Minimum coincided with a period of lower-than- average European temperatures.

Underlying variation in solar activity

Monitored sunspot activity (since 1750)

Individual records of sunspot activity (1610 onwards)

Carbon-14 proxy for solar activity (tree-ring record)

Recent eras of very low sunspot activity

Modern temperature maximum

Maunder sunspot minimum

Dalton sunspot minimum

1600   1700   1800   1900   2000   Year

2013

Northern hemisphere temperature anomaly (°C)

+0.6

Maunder Minimum

Dalton Minimum

The 'Little Ice Age'

0

Charles Dickens' childhood (1812–30): exceptionally harsh winters coloured the author's vision – the Dickensian winter of Oliver Twist.

River Thames Frost Fairs: when the river froze over in the winter

−0.6

1550   1600   1650   1700   1750   1800   1850   Year

Climate change: snowball Earth

At various times in Earth's history, it is thought that there have been periods of such intense glaciation that almost the entire planet has been covered in ice. If enough snow and ice accumulate, more and more of the Sun's energy is reflected out into space, leading to runaway cooling.

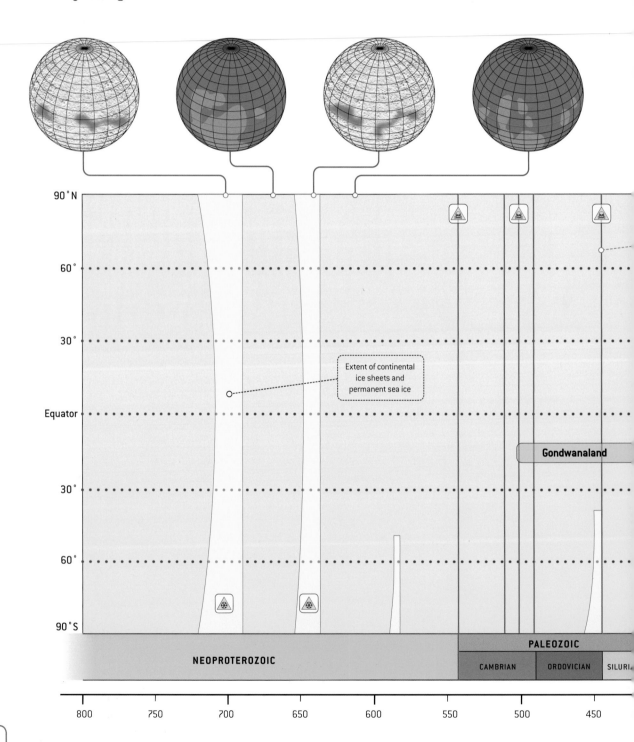

Extent of continental ice sheets and permanent sea ice

Gondwanaland

90°N
60°
30°
Equator
30°
60°
90°S

PALEOZOIC
NEOPROTEROZOIC
CAMBRIAN   ORDOVICIAN   SILURIA

800   750   700   650   600   550   500   450

MARS AND VENUS: RUNAWAY GREENHOUSE EFFECT

Just as positive feedback can lead to runaway cooling, so it can push climate in the other direction. A runaway greenhouse effect involving carbon dioxide and water vapour may have occurred on Venus until its oceans boiled away.

> Mars lost most of its atmosphere billions of years ago: low gravity and loss of its protective magnetic field contributed to the stripping away.

Atmosphere of Venus
97% carbon dioxide
Pressure 9,200% of Earth's
Average surface temp. +467 °C

Atmosphere of Earth
0.03% carbon dioxide
Pressure 100% of Earth's
Average surface temp. +15°C

Atmosphere of Mars
96% carbon dioxide
Pressure 1% of Earth's
Average surface temp. −50°C

 Snowball Earth

 Mass extinction

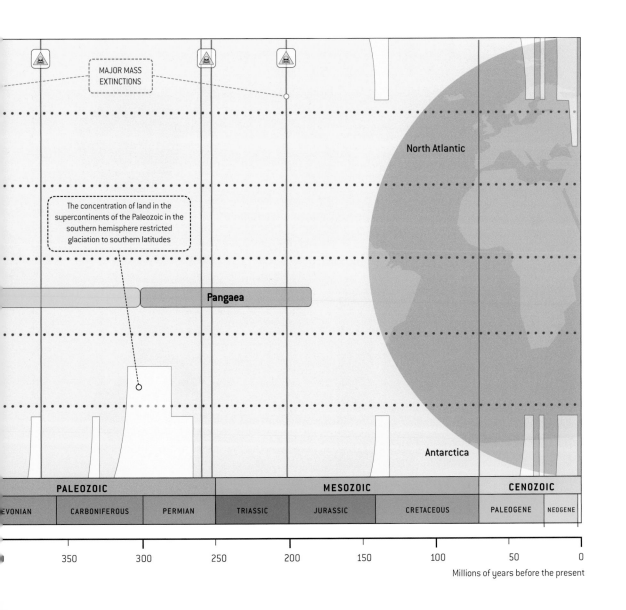

MAJOR MASS
EXTINCTIONS

North Atlantic

The concentration of land in the supercontinents of the Paleozoic in the southern hemisphere restricted glaciation to southern latitudes

Pangaea

Antarctica

PALEOZOIC			MESOZOIC			CENOZOIC	
DEVONIAN	CARBONIFEROUS	PERMIAN	TRIASSIC	JURASSIC	CRETACEOUS	PALEOGENE	NEOGENE

350   300   250   200   150   100   50   0

Millions of years before the present

Water and aqueous chemistry

Dihydrogen monoxide, otherwise known as water, is central to life. A small molecule with extraordinary properties, it is the 'goldilocks' compound: delicately balanced in its reactivity and stability.

WATER'S CHEMICAL PROPERTIES

- **Universal solvent**: dissolves more substances than any other common liquid
- **pH**: water dissociates into charged anions (OH^-) and cations (H^+)
- **Polarity**: water molecules have a positive and a negative end
- **Hydrophobic effect**: describes the segregation and apparent repulsion between water and nonpolar substances

HYDROGEN BONDING

Because water is a polarised molecule with positively and negatively charged areas, it is electrostatically attracted to other water molecules. This hydrogen bonding is responsible for water's high boiling point and surface tension.
These hydrogen-bond attractions can occur between molecules (as with water) or within different parts of a single molecule (with proteins and DNA). Hydrogen bonding is partly responsible for the secondary and tertiary structures of proteins and nucleic acids.

WATER'S PHYSICAL PROPERTIES

- Found in all three phase states (solid; liquid; gas) on Earth
- Very high surface tension
- Highest heat conduction of any liquid, other than mercury
- Highest heat capacity of all common solids and liquids
- low viscosity
- maximum density at 4 °C

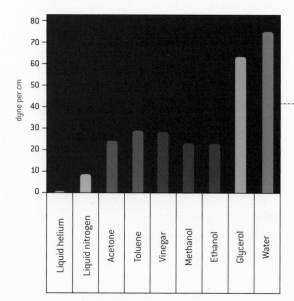

SURFACE TENSION OF COMMON SOLVENTS AND LIQUIDS

dyne per cm

	80
	70
	60
	50
	40
	30
	20
	10
	0

Liquid helium · Liquid nitrogen · Acetone · Toluene · Vinegar · Methanol · Ethanol · Glycerol · Water

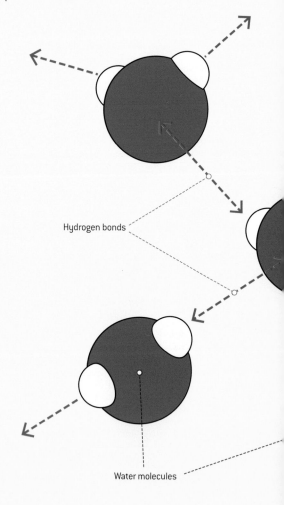

Hydrogen bonds

Water molecules

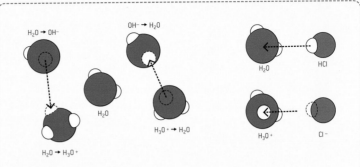

$H_2O \rightarrow OH^-$

$OH^- \rightarrow H_2O$

H_2O

H_2O

HCl

$H_3O^+ \rightarrow H_2O$

H_3O^+

Cl^-

$H_2O \rightarrow H_3O^+$

IONISATION

The self-ionisation of water is an ionisation reaction in pure water or an aqueous solution, in which a water molecule, H_2O, loses the nucleus of one of its hydrogen atoms to become a hydroxide ion, OH^-. The hydrogen nucleus, H^+, immediately protonates another water molecule to form hydronium, H_3O^+. When acids are added to water the proportion of hydronium ions increases and the pH of the solution decreases, creating an acidic solution.

WATER AS A SOLVENT: DISSOLVING IONIC SALTS

When you place an ionic substance in water, the water molecules attract the positive and negative ions from the crystal. The slightly positive hydrogen atoms are attracted to the negative chlorine **anions**; the slightly negative oxygen is attracted to the positive sodium **cations**. The salt ions, surrounded with water molecules are now free to break away and move freely in solution.

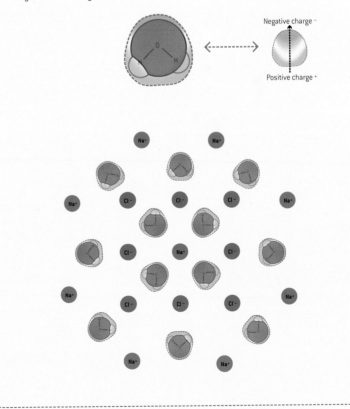

Negative charge −

Positive charge +

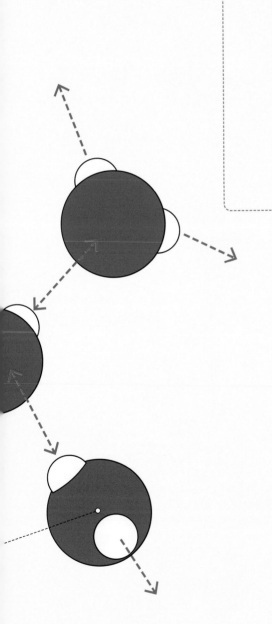

Organic chemistry

The unique properties of carbon allow it to form large molecules with long chains of carbon atoms and a variety of other elements. The resulting componds have a wide range of characteristics. The organic chemistry of life is greatly influenced by the tendency of organic molecules, or specific parts of those molecules, to attract or repel water.

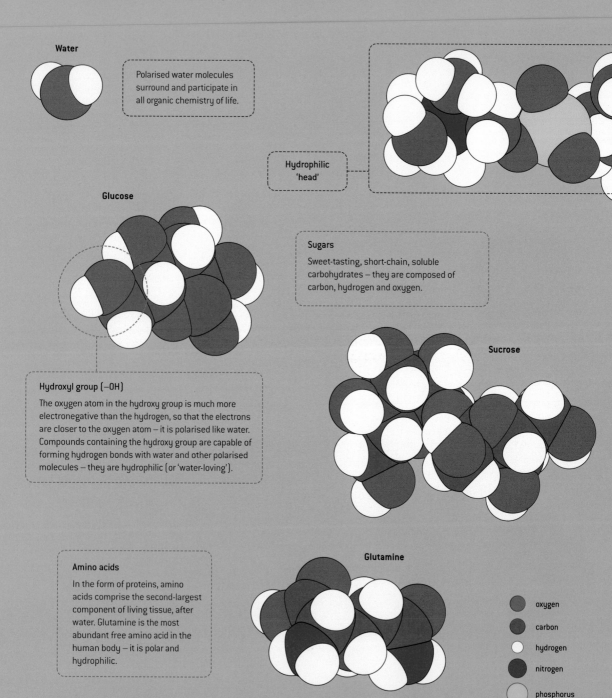

Water

Polarised water molecules surround and participate in all organic chemistry of life.

Hydrophilic 'head'

Glucose

Sugars

Sweet-tasting, short-chain, soluble carbohydrates – they are composed of carbon, hydrogen and oxygen.

Hydroxyl group (–OH)

The oxygen atom in the hydroxy group is much more electronegative than the hydrogen, so that the electrons are closer to the oxygen atom – it is polarised like water. Compounds containing the hydroxy group are capable of forming hydrogen bonds with water and other polarised molecules – they are hydrophilic (or 'water-loving').

Sucrose

Amino acids

In the form of proteins, amino acids comprise the second-largest component of living tissue, after water. Glutamine is the most abundant free amino acid in the human body – it is polar and hydrophilic.

Glutamine

- oxygen
- carbon
- hydrogen
- nitrogen
- phosphorus

Hydroxyl group
hydrophilic

Cholesterol

Cholesterol

A lipid molecule and an essential structural component of all animal cell membranes that maintains both membrane structural integrity and fluidity. The hydroxy group on cholesterol interacts with the polar head groups of the membrane phospholipids, while the bulky hydrocarbon part of the molecule is embedded in the membrane, alongside the nonpolar fatty-acid chains of the other lipids.

Phospholipid

Fatty acids

These are the main constituents of fats and oils, in the form of triglycerides, three fatty acid chains linked by a glycerol molecule. They are a rich source of energy in the body. Triglycerides of oleic acid compose the majority of olive oil and many other naturally occurring vegetable oils.

Oleic acid

Hydrophobic (or 'water-fearing'),
non-polar carbon chains
Comprised of just carbon and hydrogen.

Phospholipids

A class of lipids that are a major component of all cell membranes. They can form lipid bilayers because of their amphiphilic characteristic – both water attracting and water repelling. The structure of the phospholipid molecule generally consists of two hydrophobic fatty acid 'tails' and a hydrophilic phosphate 'head', joined together by a glycerol molecule.

Proteins

The heavy lifting of life is done by proteins and they account for around 20 per cent of our mass; structural proteins give us form and substance; as enzymes they are the catalytic foundation of all life processes. Proteins can twist and fold into a wild diversity of shapes, and can do just about anything.

Each amino acid has the same fundamental structure, with a carbon–nitrogen backbone and varying side chains that have different chemical properties: some are charged, some are polar and others are hydrophobic. In an aqueous environment the side chains seek to bond with other side chains, or bond with the watery environment … hydrophobic groups will seek to hide themselves within the folded protein.

PROPERTIES OF AMINO ACIDS

The properties of amino acids are determined by the chemical properties of their side chains

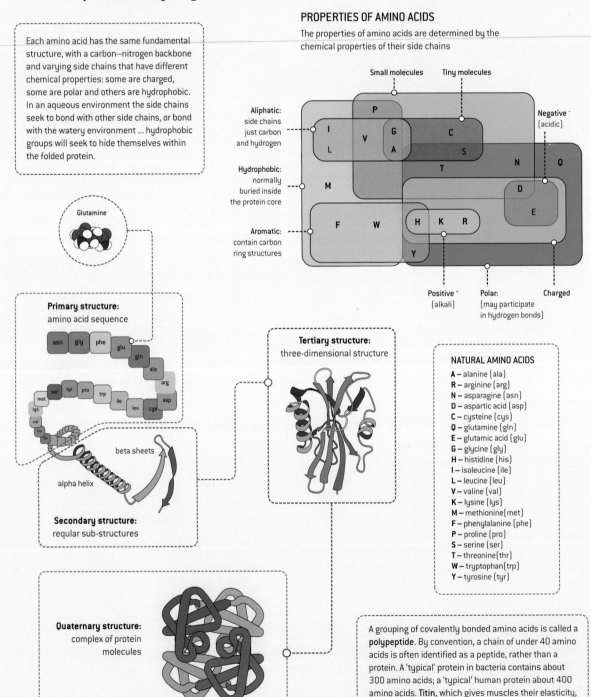

Small molecules
Tiny molecules

Aliphatic:
side chains
just carbon
and hydrogen

Hydrophobic:
normally
buried inside
the protein core

Aromatic:
contain carbon
ring structures

Negative¯
(acidic)

Positive⁺
(alkali)

Polar:
(may participate
in hydrogen bonds)

Charged

P I V G L A C S T N Q M D E F W H K R Y

Glutamine

Primary structure:
amino acid sequence

asn gly phe glu gln ala arg ser tyr pro trp ile leu asp met lys cys val his

beta sheets

alpha helix

Secondary structure:
regular sub-structures

Tertiary structure:
three-dimensional structure

NATURAL AMINO ACIDS

A – alanine (ala)
R – arginine (arg)
N – asparagine (asn)
D – aspartic acid (asp)
C – cysteine (cys)
Q – glutamine (gln)
E – glutamic acid (glu)
G – glycine (gly)
H – histidine (his)
I – isoleucine (ile)
L – leucine (leu)
V – valine (val)
K – lysine (lys)
M – methionine(met)
F – phenylalanine (phe)
P – proline (pro)
S – serine (ser)
T – threonine(thr)
W – tryptophan(trp)
Y – tyrosine (tyr)

Quaternary structure:
complex of protein molecules

A grouping of covalently bonded amino acids is called a **polypeptide**. By convention, a chain of under 40 amino acids is often identified as a peptide, rather than a protein. A 'typical' protein in bacteria contains about 300 amino acids; a 'typical' human protein about 400 amino acids. **Titin**, which gives muscles their elasticity, is a whopper, consisting of 33,423 amino acids!

UNFOLDED PROTEIN CHAIN

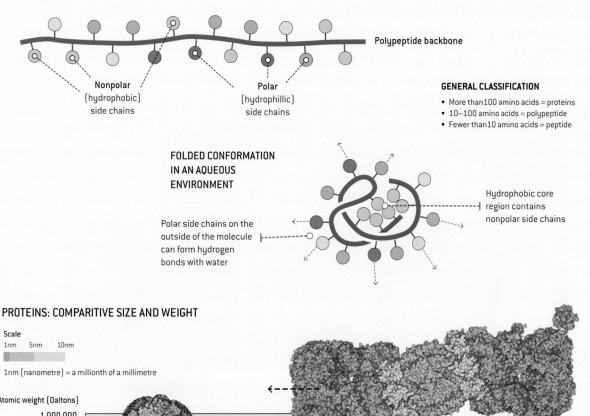

Polypeptide backbone

Nonpolar
(hydrophobic)
side chains

Polar
(hydrophillic)
side chains

GENERAL CLASSIFICATION

- More than 100 amino acids = proteins
- 10–100 amino acids = polypeptide
- Fewer than 10 amino acids = peptide

FOLDED CONFORMATION
IN AN AQUEOUS
ENVIRONMENT

Polar side chains on the
outside of the molecule
can form hydrogen
bonds with water

Hydrophobic core
region contains
nonpolar side chains

PROTEINS: COMPARITIVE SIZE AND WEIGHT

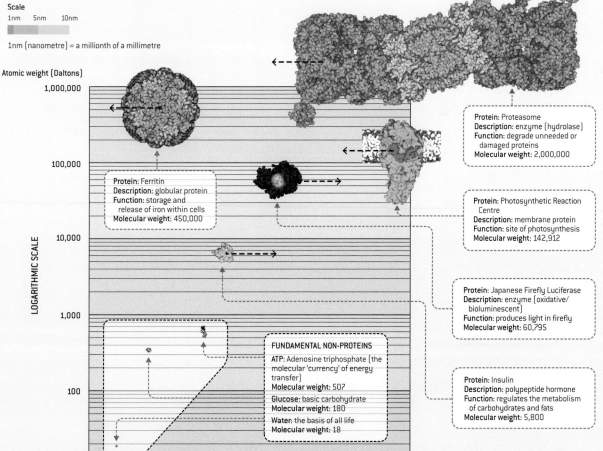

Scale

1nm   5nm   10nm

1nm (nanometre) = a millionth of a millimetre

Atomic weight (Daltons)

1,000,000

100,000

10,000

LOGARITHMIC SCALE

1,000

100

10

Protein: Proteasome
Description: enzyme (hydrolase)
Function: degrade unneeded or
    damaged proteins
Molecular weight: 2,000,000

Protein: Ferritin
Description: globular protein
Function: storage and
    release of iron within cells
Molecular weight: 450,000

Protein: Photosynthetic Reaction
    Centre
Description: membrane protein
Function: site of photosynthesis
Molecular weight: 142,912

Protein: Japanese Firefly Luciferase
Description: enzyme (oxidative/
    bioluminescent)
Function: produces light in firefly
Molecular weight: 60,795

FUNDAMENTAL NON-PROTEINS

ATP: Adenosine triphosphate (the
molecular 'currency' of energy
transfer)
Molecular weight: 507

Glucose: basic carbohydrate
Molecular weight: 180

Water: the basis of all life
Molecular weight: 18

Protein: Insulin
Description: polypeptide hormone
Function: regulates the metabolism
    of carbohydrates and fats
Molecular weight: 5,800

DNA coding

Each of the 20 amino acids that build proteins is coded for by three base pairs in the DNA sequence. These base-pair words are strung together to create specific sequences of amino acids in proteins. This is the syntax of the genome.

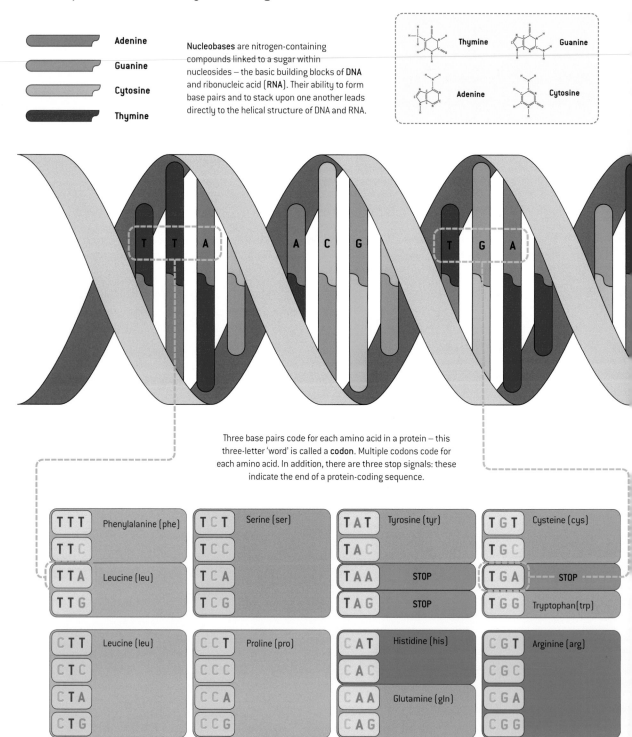

Adenine

Guanine

Cytosine

Thymine

Nucleobases are nitrogen-containing compounds linked to a sugar within nucleosides – the basic building blocks of **DNA** and ribonucleic acid (**RNA**). Their ability to form base pairs and to stack upon one another leads directly to the helical structure of DNA and RNA.

Thymine   Guanine

Adenine   Cytosine

T T A   A C G   T G A

Three base pairs code for each amino acid in a protein – this three-letter 'word' is called a **codon**. Multiple codons code for each amino acid. In addition, there are three stop signals: these indicate the end of a protein-coding sequence.

T T T	Phenylalanine (phe)
T T C	
T T A	Leucine (leu)
T T G	

T C T	Serine (ser)
T C C	
T C A	
T C G	

T A T	Tyrosine (tyr)
T A C	
T A A	STOP
T A G	STOP

T G T	Cysteine (cys)
T G C	
T G A	STOP
T G G	Tryptophan (trp)

C T T	Leucine (leu)
C T C	
C T A	
C T G	

C C T	Proline (pro)
C C C	
C C A	
C C G	

C A T	Histidine (his)
C A C	
C A A	Glutamine (gln)
C A G	

C G T	Arginine (arg)
C G C	
C G A	
C G G	

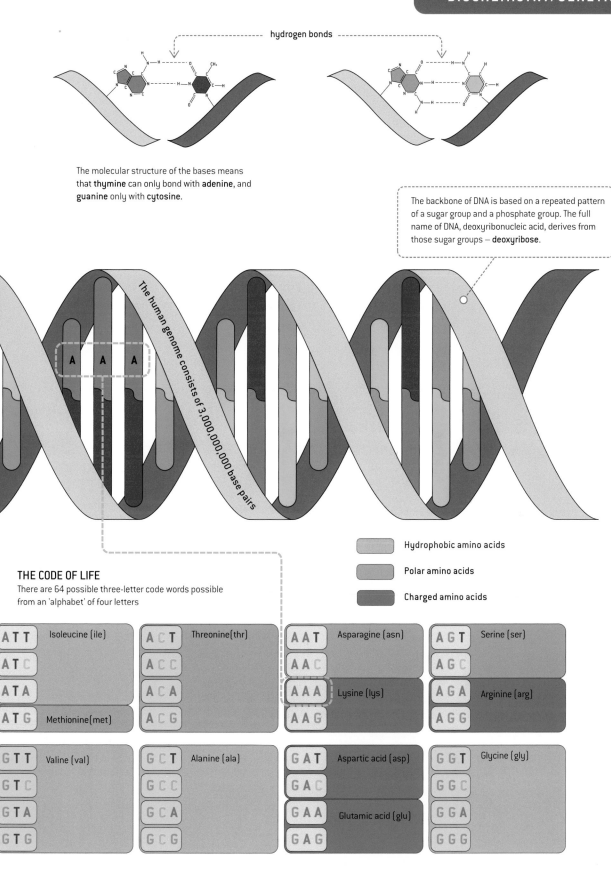

hydrogen bonds

The molecular structure of the bases means that **thymine** can only bond with **adenine**, and **guanine** only with **cytosine**.

The backbone of DNA is based on a repeated pattern of a sugar group and a phosphate group. The full name of DNA, deoxyribonucleic acid, derives from those sugar groups – **deoxyribose**.

The human genome consists of 3,000,000,000 base pairs

A A A

Hydrophobic amino acids

Polar amino acids

Charged amino acids

THE CODE OF LIFE
There are 64 possible three-letter code words possible from an 'alphabet' of four letters

A T T	Isoleucine (ile)
A T C	
A T A	
A T G	Methionine(met)

A C T	Threonine(thr)
A C C	
A C A	
A C G	

A A T	Asparagine (asn)
A A C	
A A A	Lysine (lys)
A A G	

A G T	Serine (ser)
A G C	
A G A	Arginine (arg)
A G G	

G T T	Valine (val)
G T C	
G T A	
G T G	

G C T	Alanine (ala)
G C C	
G C A	
G C G	

G A T	Aspartic acid (asp)
G A C	
G A A	Glutamic acid (glu)
G A G	

G G T	Glycine (gly)
G G C	
G G A	
G G G	

Elemental composition of crust, mantle and core of Earth (for key, see page 84)

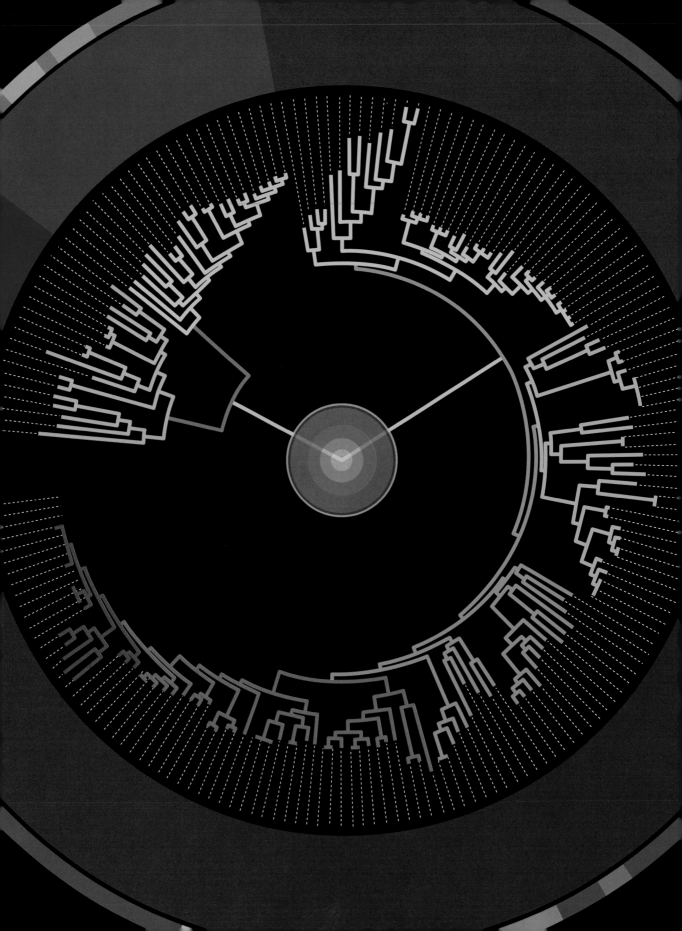

LIFE

Life

In the hot, dark, mineral-rich waters surrounding hydrothermal vents on the floor of an ancient ocean around four billion years ago, everything began to change. New chemicals appeared – sugars, nucleic acids, fatty acids, amino acids. Some of these compounds had catalytic qualities: they assisted other chemical reactions, initially fuelled by the escaping heat of the Earth's interior. As the 'soup' became richer, larger molecules began not just to assist the reaction and creation of similar molecules, but also to copy themselves – to encode information in the arrangement of the smaller molecules that made them up. The environment favoured certain groups of self-replicating molecules that could not only catalyse their own creation, but could also catalyse the creation of proteins whose structures echoed their own peculiar structure. The link between the information contained in molecules and the production of particular proteins was forged. The chemistry of life as we know it was born. And when that chemistry became contained within a membrane, life itself was born: self-replicating entities with a history, the ability to draw energy from the environment, control their internal environment, and reproduce into a variable future.

For at least one and a half billion years life continued to evolve in a world devoid of oxygen. Life's gyre was disconnected from the ever-growing power of the Sun's energy beating down on the barren lands and empty shores. Eventually, though, when remains unclear, in the shallower waters around the emerging continents the next revolution of life began. Cyanobacteria evolved with the ability to harness the power of the Sun's rays to split water, harvesting energetic electrons to turn carbon dioxide and water into sugar and oxygen.

The oxygen produced by this new form of life took another billion years to neutralise the ancient anoxic oceans and begin to fill the atmosphere. Around this time the next huge step in life's progress occurred. Certain bacteria were swallowed and began living within other bacteria; they became subservient, symbiotic power stations, allowing the host cells to blossom in size and complexity. Eukaryote cells were born. These were cells with a nucleus and organelles that could reproduce sexually and went on to evolve into multicellular organisms: life bigger than a pinprick. Another billion years of slow-ticking progress ...

Six hundred million years before the present day, the first complex fossils appear in the record: the Ediacaran Biota – strange and mysterious beasts oozing around the silty floors of shallow seas. Another fifty million years and the dawn of the Cambrian age: life explodes in form and variety. Within 25 million years nearly all existing groups of modern animals had emerged. Meanwhile the levels of oxygen in the atmosphere continued to grow; on high, the Sun's radiation converted oxygen to ozone, creating a protective blanket to absorb the harshest ultraviolet light. But, apart from the odd mollusc or arthropod making brief forays onto the damp shoreline, life remained trapped within the protective incubator of the seas.

To conquer the land, to breathe the air! But how to avoid a rapid desiccated demise? How to bear young when all life needs water? The obstacles were great, but the rewards were legion. First plants, then animals – insects and molluscs, followed 50 million years later by the first land vertebrates – dragged themselves up onto the land. Transporting and preserving water, armouring themselves against the dry air, learning how to breathe – wood and bones defying the loss of water's buoyancy; seeds and eggs coddling the young.

The final 350 million years of our history is more familiar: towering forests, huge insects, animals taking to the skies, giant reptiles, mass extinctions, ice ages, warm blood, vast continents breaking apart, the worldwide spread of open grasslands, the rise of mammals, the age of the ape, and – for better or worse – the emergence of the self-reflective sentience of humankind.

Abiogenesis: the start of life

Abiogenesis is the process by which life might have arisen naturally from non-living matter. The ultimate chicken-and-egg story ...

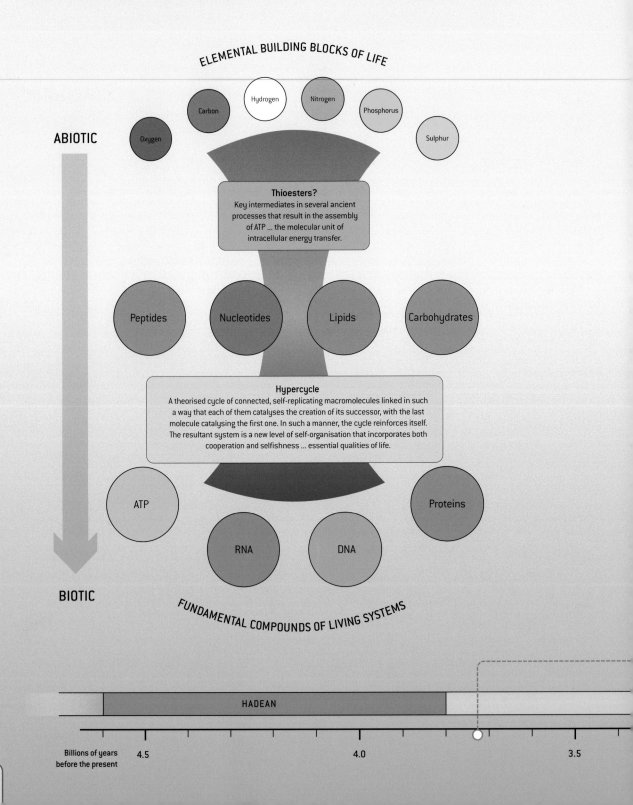

ELEMENTAL BUILDING BLOCKS OF LIFE

Oxygen
Carbon
Hydrogen
Nitrogen
Phosphorus
Sulphur

ABIOTIC

Thioesters?
Key intermediates in several ancient processes that result in the assembly of ATP ... the molecular unit of intracellular energy transfer.

Peptides
Nucleotides
Lipids
Carbohydrates

Hypercycle
A theorised cycle of connected, self-replicating macromolecules linked in such a way that each of them catalyses the creation of its successor, with the last molecule catalysing the first one. In such a manner, the cycle reinforces itself. The resultant system is a new level of self-organisation that incorporates both cooperation and selfishness ... essential qualities of life.

ATP
Proteins
RNA
DNA

BIOTIC

FUNDAMENTAL COMPOUNDS OF LIVING SYSTEMS

HADEAN

Billions of years before the present

4.5

4.0

3.5

LOST CITY HYDROTHERMAL FIELD

Hydrothermal vents are fissures in a planet's lithosphere from which geothermally heated water issues. They are commonly found where tectonic plates are moving apart at mid-ocean ridges. Vents on land lead to hot springs, fumaroles and geysers. In 2000, an unusual, highly alkaline field of vents was discovered in the middle of the Atlantic Ocean: the Lost City hydrothermal field consists of about thirty chimneys made of calcium carbonate 30–60 metres tall. The Lost City provides a working model for the study of the origins of life and other processes driven by the abiotic production of methane and hydrogen by serpentinisation. Hydrogen-saturated alkaline water meeting acidic oceanic water would produce a **natural proton gradient** across thin mineral 'walls' in the vents that are rich in catalytic iron–sulphur minerals. This could create the right conditions for converting carbon dioxide and hydrogen into organic molecules, which can then react with each other to form the building blocks of life such as nucleotides and amino acids.

ACETATE

In nature, acetate is the most common building block for biosynthesis and lies at the heart of biochemistry. For instance, fatty acids are produced by connecting the two carbon atoms from acetate to a growing carbon chain. Acetate is principally utilised by organisms in the form of acetyl coenzyme A – an important molecule in metabolism, used in many biochemical reactions. Its main function is to convey carbon atoms to the citric acid cycle which lies at the heart of cell metabolism.

THIOESTERS

Could be the key group of chemicals involved at the dawn of life. They participate in the synthesis of a number of other cellular components, including peptides, fatty acids, sterols, terpenes and porphyrins.

SERPENTINISATION

A geological low-temperature metamorphic process involving heat and water in which rocks are oxidised (anaerobic oxidation of iron by the protons of water leading to the formation of H_2) and hydrolysed with water into serpentine. The reactions create a highly alkaline environment.

CH_3COO^- (Acetate)

NO

BIOSYNTHESIS?

BIOTIC?

ABIOTIC

CH_3COSCH_3

H^+

$\cdot CH_3$

CO

NO_2

CO_2

OCEAN FLOOR

H_2

Uprising, hot, alkali water about 100˚C/pH 10.5

H_2

CH_4

SERPENTINISED HIGHLY ALKALINE ROCKS

Nitric oxide is a key biological messenger in vertebrates, playing a role in a variety of biological processes

Nitrites

Abiotic methane

Serpentinisation has been proposed as the process that produces the methane detected on the surface of Mars and in the liquid water ocean hidden beneath the frozen surface of Saturn's moon, Enceladus.

FIRST LIFE?

ARCHEAN

3.0   2.5

First life: chemistry changes

The ingredients of life were always here in the stardust that formed out planet, but the birth of life radically changed the chemistry and elemental constituents that existed in the air, the earth and the oceans.

ELEMENTAL COMPOSITION OF THE ATMOSPHERE

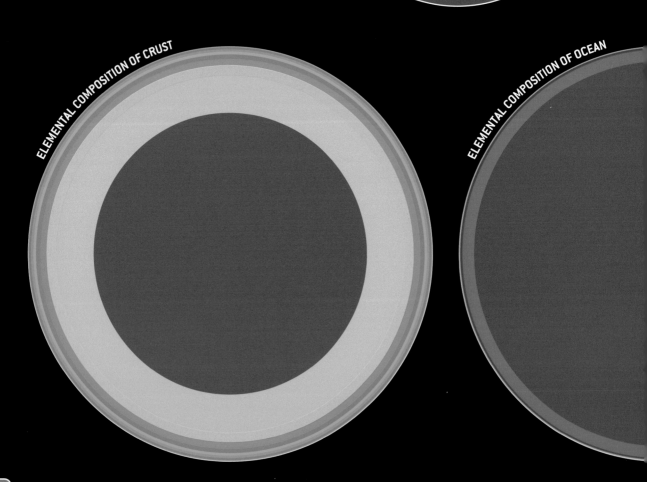

ELEMENTAL COMPOSITION OF CRUST

ELEMENTAL COMPOSITION OF OCEAN

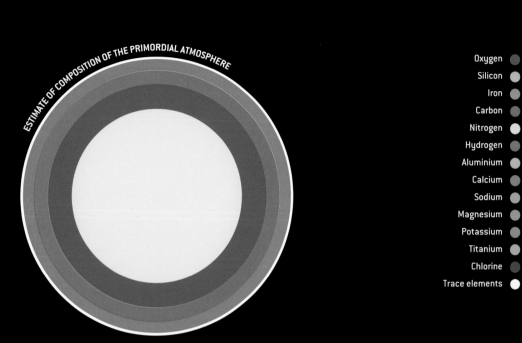

ESTIMATE OF COMPOSITION OF THE PRIMORDIAL ATMOSPHERE

Oxygen	
Silicon	
Iron	
Carbon	
Nitrogen	
Hydrogen	
Aluminium	
Calcium	
Sodium	
Magnesium	
Potassium	
Titanium	
Chlorine	
Trace elements	

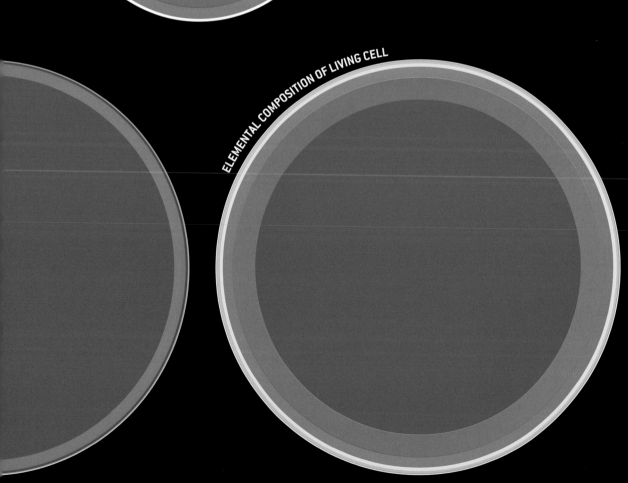

ELEMENTAL COMPOSITION OF LIVING CELL

Protocell: the first cell

The earliest forms of life needed a membrane for
many of the same reasons that modern cells do:
to keep molecules that are important for growth
and survival accessible and in high enough
concentration and to keep unneeded or
potentially harmful molecules out of the cell.

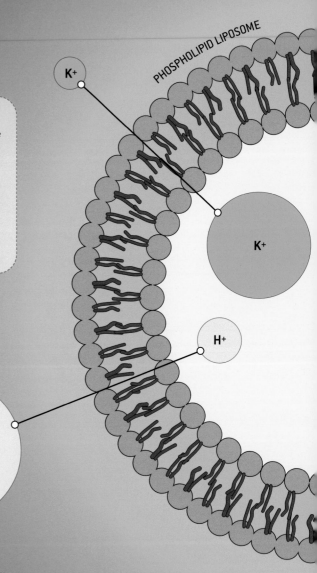

PHOSPHOLIPID LIPOSOME

NATURAL PROTON GRADIENTS

Four billion years ago, alkaline fluids bubbled into what would then have
been mildly acidic oceans. Atmospheric CO_2 levels were about a
thousand times higher than they are today, and CO_2 forms carbonic
acid when it disolves in water. Acidity is just a measure of proton
concentration, which was about 10,000 time higher (four pH units) in
the oceans than in vent fluids. That difference gave rise to a natural
proton gradient across the vent membranes that had the same polarity
(outside positive) and a similar electrochemical potential as modern
cells have. Thermodynamic arguments suggest that the only way life
could have started at all is if it found a way to tap this proton gradient.

MEMBRANE POTENTIAL

A typical animal cell has a transmembrane electrical potential
between -50 millivolts and -100 millivolts inside the cell ... up to about
one-tenth of a volt. This voltage gradient allows a cell to function like a
battery, providing power to operate the various protein machines (ion
pumps, motors, transporters) embedded in the membrane.

CELLS PROVIDE TWO FUNDAMENTAL REQUIREMENTS FOR LIFE

- **protection from the outside environment**: keeping complex molecules stable in a varying and sometimes hostile environment
- **confinement of biochemical activity**: required for the evolution of biocomplexity. If the molecules that code for enzymes (DNA and RNA) are not enclosed in cells, the enzymes evolved by a particular group of replicators will automatically benefit the neighbouring replicator molecules. The consequences of such genetic diffusion in theoretical non-partitioned life forms can be viewed as parasitism by default. Selection pressures on replicator molecules would be lower, as the molecules that produces the better enzymes have no definitive advantage over their close neighbours. If biochemistry is enclosed by a cell membrane, then well-coded enzymes will only be available to the replicator molecule itself. That molecule will uniquely benefit from the enzymes it codes for, giving it a better chance to multiply

Cl −

Cl −

Na+

Na+

Phospholipid bilayers are relatively impermeable to molecules such as nucleotides and require special embedded transporter proteins to allow their passage through the membrane.

Vesicle formation may be catalysed by certain clays, which have also been found to catalyse the formation of strands of RNA from single nucleotides – clays such as **montmorillonite** may very well have been the key in the formation of the first protocells. These minerals can catalyse the stepwise formation of the hydrocarbon tails of fatty acids from hydrogen and carbon monoxide gases released from hydrothermal vents. Fatty acids of various lengths are eventually released into the surrounding water. Similar to phospholipids, fatty acids have a hydrophobic tail and hydrophilic head, and can thus form the same types of structures, such as bilayers, vesicles and micelles. Fatty acid liposomes would protect the internal environment of the protocell, while allowing for some inward and outward movement of the essential molecules of life.

Cell membranes

Cell membranes separate the interior of all cells from the outside environment, controlling the movement of substances in and out of cells, protecting the cell from its surroundings. It consists of a phospholipid bilayer and embedded proteins.

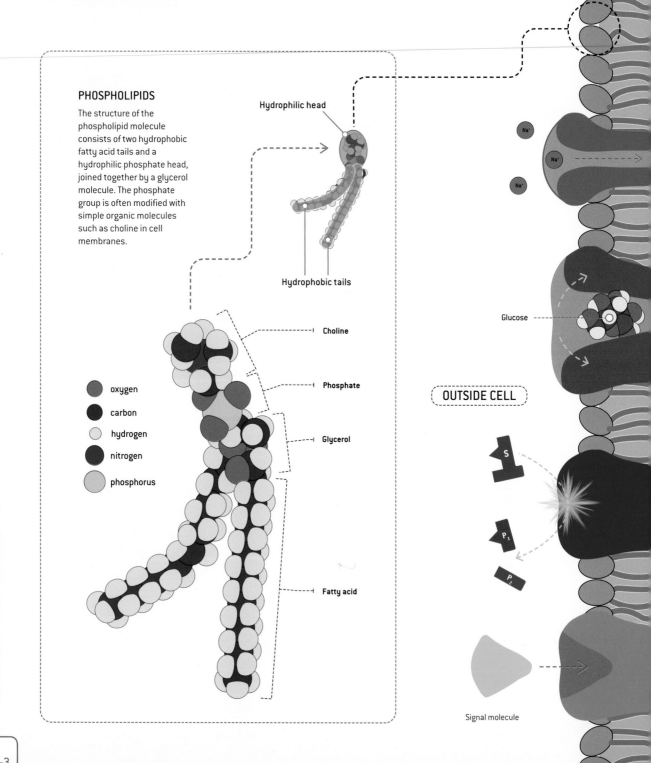

PHOSPHOLIPIDS

The structure of the phospholipid molecule consists of two hydrophobic fatty acid tails and a hydrophilic phosphate head, joined together by a glycerol molecule. The phosphate group is often modified with simple organic molecules such as choline in cell membranes.

Hydrophilic head

Hydrophobic tails

Choline

Phosphate

Glycerol

Fatty acid

- oxygen
- carbon
- hydrogen
- nitrogen
- phosphorus

Na^+

Na^+

Na^+

Glucose

OUTSIDE CELL

S

P_1

P_2

Signal molecule

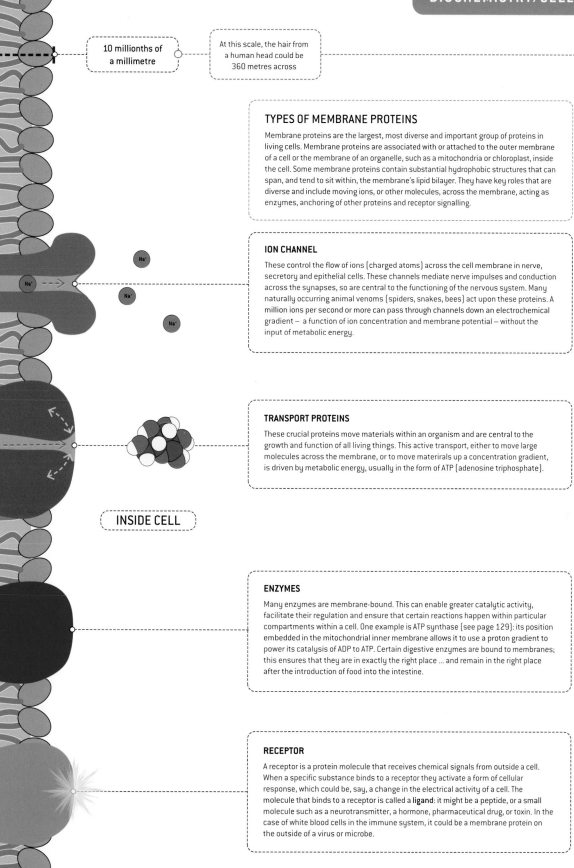

10 millionths of
a millimetre

At this scale, the hair from
a human head could be
360 metres across

TYPES OF MEMBRANE PROTEINS

Membrane proteins are the largest, most diverse and important group of proteins in living cells. Membrane proteins are associated with or attached to the outer membrane of a cell or the membrane of an organelle, such as a mitochondria or chloroplast, inside the cell. Some membrane proteins contain substantial hydrophobic structures that can span, and tend to sit within, the membrane's lipid bilayer. They have key roles that are diverse and include moving ions, or other molecules, across the membrane, acting as enzymes, anchoring of other proteins and receptor signalling.

ION CHANNEL

These control the flow of ions (charged atoms) across the cell membrane in nerve, secretory and epithelial cells. These channels mediate nerve impulses and conduction across the synapses, so are central to the functioning of the nervous system. Many naturally occurring animal venoms (spiders, snakes, bees) act upon these proteins. A million ions per second or more can pass through channels down an electrochemical gradient – a function of ion concentration and membrane potential – without the input of metabolic energy.

TRANSPORT PROTEINS

These crucial proteins move materials within an organism and are central to the growth and function of all living things. This active transport, either to move large molecules across the membrane, or to move materirals up a concentration gradient, is driven by metabolic energy, usually in the form of ATP (adenosine triphosphate).

INSIDE CELL

ENZYMES

Many enzymes are membrane-bound. This can enable greater catalytic activity, facilitate their regulation and ensure that certain reactions happen within particular compartments within a cell. One example is ATP synthase (see page 129): its position embedded in the mitochondrial inner membrane allows it to use a proton gradient to power its catalysis of ADP to ATP. Certain digestive enzymes are bound to membranes; this ensures that they are in exactly the right place ... and remain in the right place after the introduction of food into the intestine.

RECEPTOR

A receptor is a protein molecule that receives chemical signals from outside a cell. When a specific substance binds to a receptor they activate a form of cellular response, which could be, say, a change in the electrical activity of a cell. The molecule that binds to a receptor is called a **ligand**: it might be a peptide, or a small molecule such as a neurotransmitter, a hormone, pharmaceutical drug, or toxin. In the case of white blood cells in the immune system, it could be a membrane protein on the outside of a virus or microbe.

Kingdoms of life

The ranking and categorising of life helps us to understand the complexity of the natural world and its evolutionary history. But the portrait is ever-changing and dependent on how you look at it. Life fits uneasily in boxes.

THE THREE DOMAINS

The kingdoms Archaea and Bacteria are as biochemically different from each other as from eukaryotes, so they are assigned a higher category, a **domain**. This is a tree of life based on ribosomal RNA, where the length of branches represents genetic distance and colour represents the maximum temperature at which a species will grow.

Historically, evolutionary trees were all based on phylogeny – the physical, anotomical characteristics of species. Recent advances in genetic sequencing have allowed scientists to reconstruct evolutionary history by comparing the alignment of tens to hundreds of genes.

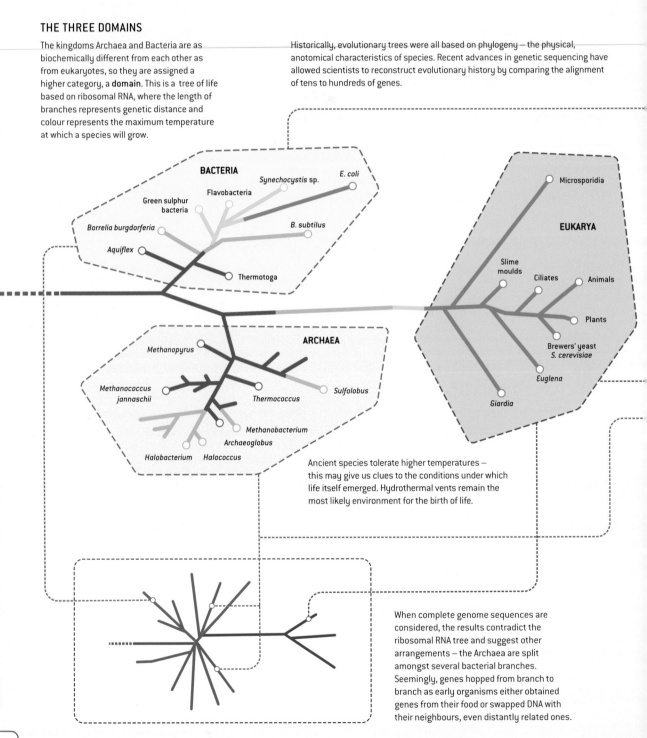

BACTERIA

Synechocystis sp.
E. coli
Flavobacteria
Green sulphur bacteria
Borrelia burgdorferia
B. subtilus
Aquiflex
Thermotoga

EUKARYA

Microsporidia
Slime moulds
Ciliates
Animals
Plants
Brewers' yeast S. cerevisiae
Euglena
Giardia

ARCHAEA

Methanopyrus
Methanococcus jannaschii
Thermococcus
Sulfolobus
Methanobacterium
Archaeoglobus
Halobacterium
Halococcus

Ancient species tolerate higher temperatures – this may give us clues to the conditions under which life itself emerged. Hydrothermal vents remain the most likely environment for the birth of life.

When complete genome sequences are considered, the results contradict the ribosomal RNA tree and suggest other arrangements – the Archaea are split amongst several bacterial branches. Seemingly, genes hopped from branch to branch as early organisms either obtained genes from their food or swapped DNA with their neighbours, even distantly related ones.

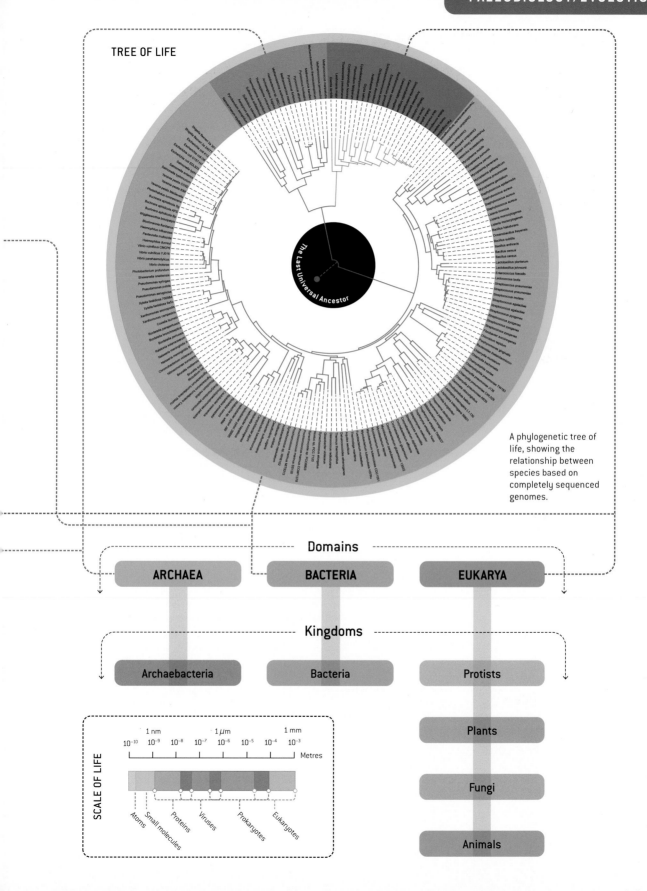

TREE OF LIFE

A phylogenetic tree of life, showing the relationship between species based on completely sequenced genomes.

Domains

ARCHAEA

BACTERIA

EUKARYA

Kingdoms

Archaebacteria

Bacteria

Protists

Plants

Fungi

Animals

SCALE OF LIFE

1 nm   1 μm   1 mm

10^{-10} 10^{-9} 10^{-8} 10^{-7} 10^{-6} 10^{-5} 10^{-4} 10^{-3}

Metres

Atoms

Small molecules

Proteins

Viruses

Prokaryotes

Eukaryotes

Eukaryotic Cells: the origins

All plants and animals are made up of eukaryotic cells: the sheer complexity and variety of these cells was made possible by the evolution of the photosynthetic and respiratory powerstations that are chloroplasts and mitochondria.

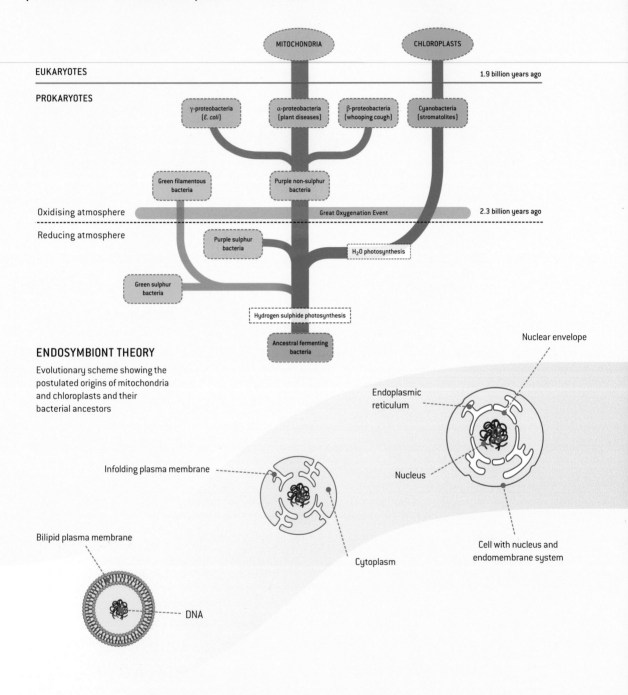

MITOCHONDRIA

CHLOROPLASTS

EUKARYOTES   1.9 billion years ago

PROKARYOTES

γ-proteobacteria
(E. coli)

α-proteobacteria
(plant diseases)

β-proteobacteria
(whooping cough)

Cyanobacteria
(stromatolites)

Green filamentous
bacteria

Purple non-sulphur
bacteria

Oxidising atmosphere   Great Oxygenation Event   2.3 billion years ago

Reducing atmosphere

Purple sulphur
bacteria

H_2O photosynthesis

Green sulphur
bacteria

Hydrogen sulphide photosynthesis

Ancestral fermenting
bacteria

ENDOSYMBIONT THEORY

Evolutionary scheme showing the
postulated origins of mitochondria
and chloroplasts and their
bacterial ancestors

Nuclear envelope

Endoplasmic
reticulum

Nucleus

Infolding plasma membrane

Cell with nucleus and
endomembrane system

Bilipid plasma membrane

Cytoplasm

DNA

ANCESTRAL PROKARYOTE

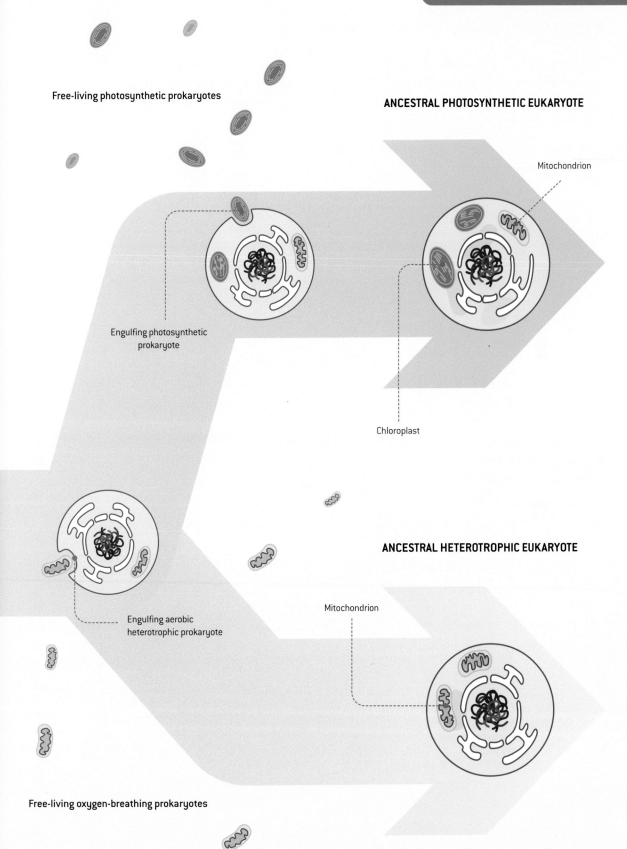

Free-living photosynthetic prokaryotes

ANCESTRAL PHOTOSYNTHETIC EUKARYOTE

Mitochondrion

Engulfing photosynthetic
prokaryote

Chloroplast

ANCESTRAL HETEROTROPHIC EUKARYOTE

Mitochondrion

Engulfing aerobic
heterotrophic prokaryote

Free-living oxygen-breathing prokaryotes

Mitochondria

Mitochondria are the powerhouses of eukaryotic cells, where the biochemical processes of respiration and energy production occur. They generate most of the cell's supply of adenosine triphosphate, which is the principal energy currency of the cell that powers thermodynamically unfavourable reactions.

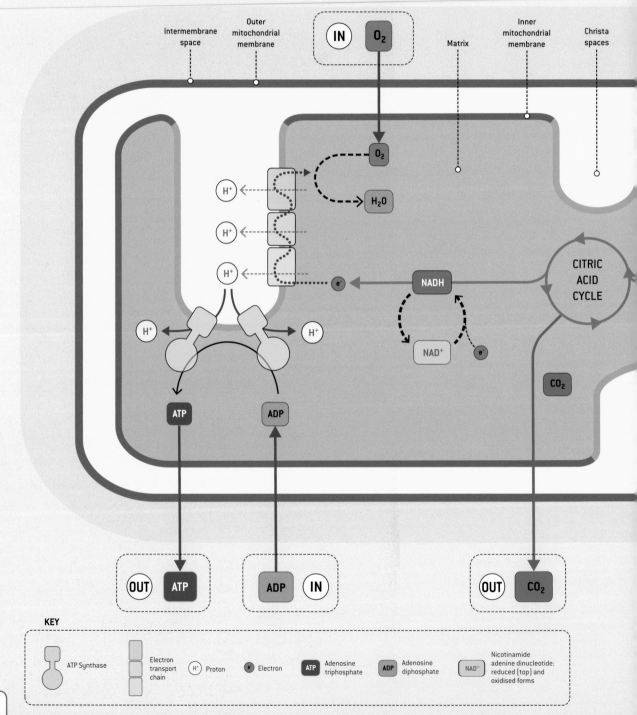

FOOD MOLECULES INSIDE THE CELL

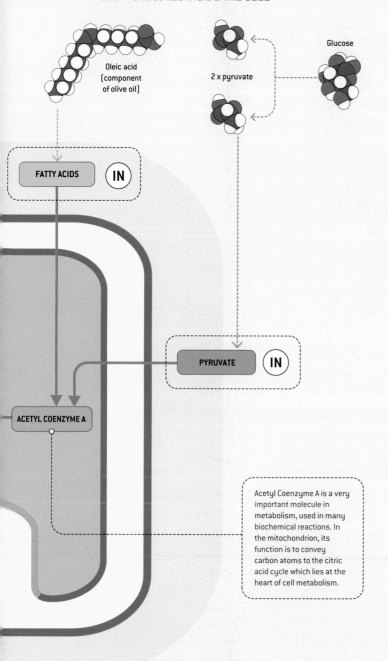

Oleic acid
(component
of olive oil)

2 x pyruvate

Glucose

FATTY ACIDS (IN)

PYRUVATE (IN)

ACETYL COENZYME A

Acetyl Coenzyme A is a very important molecule in metabolism, used in many biochemical reactions. In the mitochondrion, its function is to convey carbon atoms to the citric acid cycle which lies at the heart of cell metabolism.

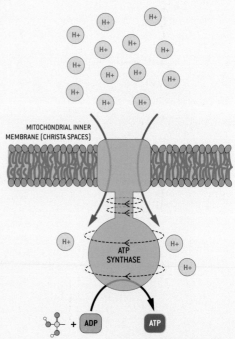

ADENOSINE TRIPHOSPHATE (ATP)

PHOSPHATE GROUP

ADENOSINE DIPHOSPHATE (ADP)

A phosphate group (PO_3) can be added to ADP by a process called **phosphorylation**, storing energy in the bond that can be released for a staggering range of metabolic processes at a later time.

H+ H+ H+ H+ H+ H+ H+ H+ H+ H+ H+ H+

MITOCHONDRIAL INNER
MEMBRANE (CHRISTA SPACES)

H+   ATP
SYNTHASE   H+

H+

+ ADP → ATP

Mitochondria vary in size from 0.5–10 micrometres; cells contain anything from 1 to 1,000 mitochondria. All mitochondria, in both plants and animals, have vary similar structures.

- there is a double membrane
- here is a space between the inner and outer membranes called the intermembrane space
- the space within the inner membrane is called the matrix
- the inner membrane is very convoluted – the inturnings are called cristae
- the protein and phospholipid composition of the inner and outer membranes are very different.

ATP SYNTHASE

Energy from oxidising food (transported via **NADH**) is used to pump **protons** (hydrogen ions) out of the matrix into the intermembrane space, using the **electron transport chain**. This creates a voltage of 170mV and a proton gradient. ATP synthase is an enzyme, an ion pump and a molecular motor in one – a true nanoscale machine. Powered by the flow of hydrogen ions back into the matrix, a circular rotor spins at 9,000 rpm, which in turn powers another motor that harnesses mechanical energy to position and catalyse the phosphorylation of ADP to the energy-charged ATP ... which is then transported away to power myriad life processs.

Viruses

The evolutionary origin of viruses is unclear, though it is likely that they evolved after bacterial life. Viruses are non-living as they exist in an inert state outside of a host cell. They consist of a strand of genetic material surrounded by a protective protein coat called a capsid, and sometimes an outer lipid envelope.

An icosahedron is a polyhedron having 20 equilateral triangular faces and 12 vertices

Fully infective viral particles can self-assemble in a test tube from purified RNA and protein molecules.

FOUR BASIC VIRAL FORMS

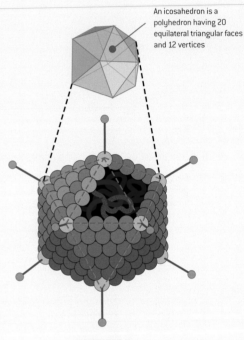

COMPLEX
Bacteriophage:
infects and kills
bacterial cells

The genetic material in a virus may take one of four forms: a double strand of DNA, a single strand of DNA, a double strand of RNA, or a single strand of RNA. The size of the genetic material of viruses is often quite small. The human genome consists of around 100,000 genes, whereas viruses typically contain between 10 and 200 genes.

POLYHEDRAL
Adenovirus:
common cause of
colds in children

HELICAL
Tobacco mosaic virus:
wide-ranging plant
virus – damages leaves

A single strand of RNA with 6,395 nucleotides, packaged in a helical coat constructed from 2,130 copies of a capsid protein, each 158 amino acids long.

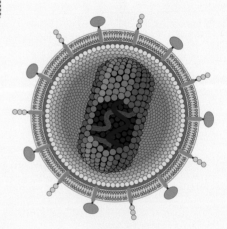

ENVELOPED
HIV virus:
attacks the human
immune system

VIRAL SCALE

E. Coli bacterium
3,000 x 800 nanometres

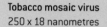

Tobacco mosaic virus
250 x 18 nanometres

Human HIV virus
130 nanometres

Human red blood cell
7,000 nanometres in diameter

Polio virus
30 nanometres

Adenovirus
75 nanometres

A nanometer is a millionth
of a millimetre

Bacteriophage T2
200 x 70 nanometres

Head of human spermatozoa
5,000 nanometres long

AT THIS SCALE:

- a human egg would be 6.5 metres in diameter;
- an average human hair would be 5 metres in diameter.

The Great Oxygenation Event

Earth's earliest atmosphere contained no free oxygen – all the oxygen that living organisms breathe today, both in the air and dissolved in the rivers and oceans, is the product of billions of years of photosynthesis.

Archaea/Bacteria

PRECAMBRIAN

PHOTOSYNTHESIS
Cyanobacteria were the first organisms to evolve the ability to photosynthesize, splitting water, fixing carbon from CO_2 using the Sun's energy, and introducing a steady supply of oxygen into the environment. For nearly a billion years, oxygen levels did not increase substantially in the atmosphere as the oxygen produced quickly reacted with iron and other minerals in the surrounding rock and ocean water.

BANDED IRON LAYERS
These were formed in sea water as the result of the released oxygen which combined with dissolved iron in Earth's anoxic and acidic oceans to form insoluble iron oxides, which precipitated out, forming a thin layer on the ocean floor.

First water-splitting photosynthesis releases O_2

Formation of oceans and continents

First living cells

First photosynthetic cells

Formation of the Earth

BANDED IRON FORMATIONS

HADEAN

ARCHEAN

4.5

40

3.5

3.0

2.5

Billions of years before the present

THE GREAT OXYGENATION EVENT

Multicellular animals

Eukaryotes

Once a saturation point was reached for the reactions in rock and water, oxygen was able to exist as a gas in its diatomic form. The concentration of oxygen in the atmosphere has risen gradually over the subsequent 2.5 billion years.

Glaciation events

Aerobic respiration
becomes widespread

First land plants

ATMOSPHERIC
OXYGEN LEVELS

First multicellular
plants and animals

BORING BILLION
Characterised by climatic stability, low levels of atmospheric oxygen,
lack of biological events, and the absence of extreme changes in the
atmospheric and oceanic composition

First vertebrates

Proterozoic glacial gap

Origin of eukaryotic
photosynthetic cells

20

10

Oxygen in atmosphere [%]

0

| PALEOPROTEROZOIC | MESOPROTEROZOIC | NEOPROTEROZOIC | PALEOZOIC | MESOZOIC |

2.0   1.5   1.0   0.5   0

Photosynthesis and primary production

Primary production is the synthesis of organic compounds from atmospheric or aqueous carbon dioxide, principally through the process of photosynthesis. Almost all life on Earth relies directly or indirectly on primary production.

The upper atmosphere of Earth receives an estimated 174,000 terawatts of incoming solar radiation. About a third is reflected back into space while the rest is absorbed by clouds, oceans and land masses. Most humans live in areas where incoming solar energy approximates to 150 to 300 watts per square metre ... when the Sun shines!

There are several types of chlorophyll with different side chains, but all share the magnesium containing **chlorin** ring (shaded green in this diagram). It's similar in structure to the **haem** structure that contains an iron atom in haemoglobin molecules.

Green plants are green because chlorophyll absorbs certain wavelengths of light within the visible light spectrum. The absorption spectrum below shows chlorophyll absorbing light in the red (long wavelength) and the blue (short wavelength) regions of the visible light spectrum. Green light is not absorbed but reflected, making the plant appear green.

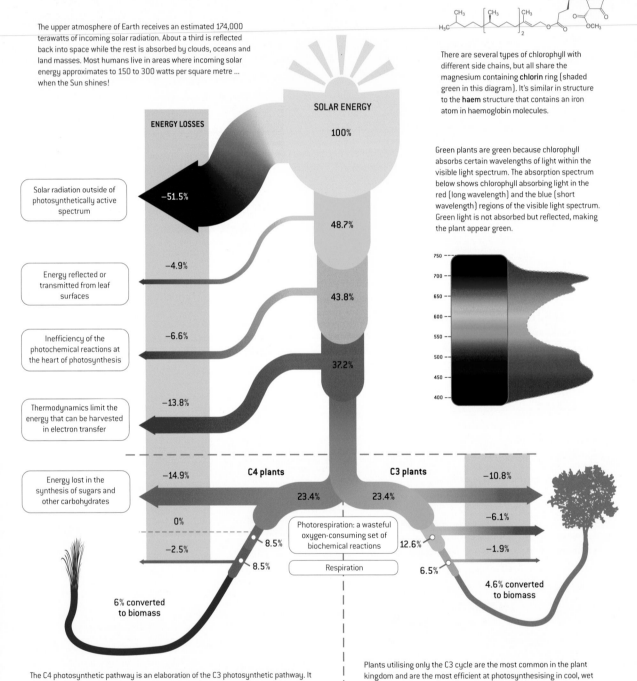

SOLAR ENERGY
100%

ENERGY LOSSES

Solar radiation outside of photosynthetically active spectrum
−51.5%

48.7%

Energy reflected or transmitted from leaf surfaces
−4.9%

43.8%

Inefficiency of the photochemical reactions at the heart of photosynthesis
−6.6%

37.2%

Thermodynamics limit the energy that can be harvested in electron transfer
−13.8%

Energy lost in the synthesis of sugars and other carbohydrates
−14.9%

C4 plants   **C3 plants**
23.4%   23.4%
−10.8%

0%

Photorespiration: a wasteful oxygen-consuming set of biochemical reactions

−6.1%

−2.5%   8.5%   12.6%   −1.9%

Respiration
8.5%   6.5%

6% converted to biomass

4.6% converted to biomass

The C4 photosynthetic pathway is an elaboration of the C3 photosynthetic pathway. It evolved as an adaptation to high light intensities, high temperatures, and dryness. C4 plants dominate grassland floras and biomass production in the warmer climates of the tropical and subtropical regions. They account for only 3% of land plant species, yet those species are responsible for almost a quarter of photosynthetic output.

Plants utilising only the C3 cycle are the most common in the plant kingdom and are the most efficient at photosynthesising in cool, wet climates. They comprise about 85% of all plant species including evergreen trees, deciduous trees and important crop plants like rice, barley and soybean. Many plants that live in the tropics and subtropics, where precipitation is high, are C3 plants.

WORLD NET PRIMARY PRODUCTION AND BIOMASS
FOR VARIOUS ECOSYSTEM TYPES

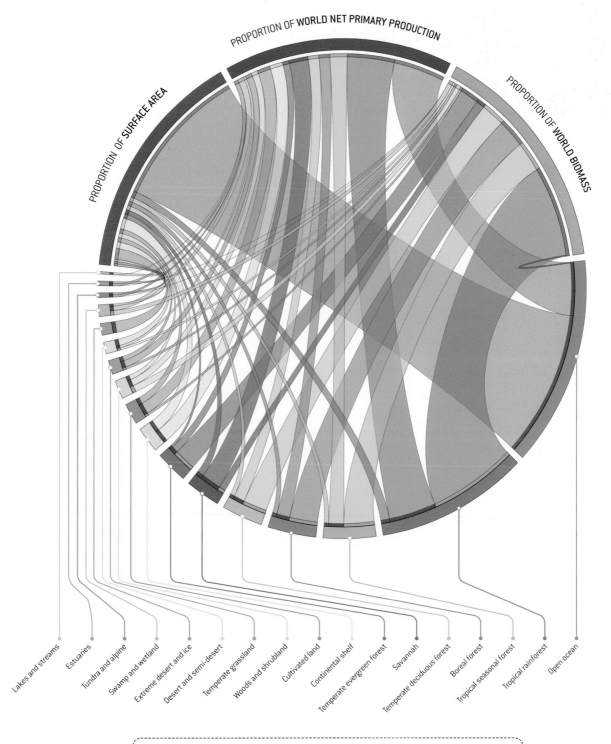

PROPORTION OF SURFACE AREA

PROPORTION OF WORLD NET PRIMARY PRODUCTION

PROPORTION OF WORLD BIOMASS

Lakes and streams
Estuaries
Tundra and alpine
Swamp and wetland
Extreme desert and ice
Desert and semi-desert
Temperate grassland
Woods and shrubland
Cultivated land
Continental shelf
Temperate evergreen forest
Savannah
Temperate deciduous forest
Boreal forest
Tropical seasonal forest
Tropical rainforest
Open ocean

The average rate at which energy is captured on Earth by photosynthesis is about 130 terawatts – equivalent to three times the power consumption of human civilisation. Photosynthetic organisms convert around 100–115 billion tonnes of carbon into biomass each year.

Energy consumption of
human brain
46 watts

(2% of body mass, but 20%
of energy consumption)

ENERGY

Resting metabolic
rate in average
human
115 watts

Maximum
mechanical
power output by
professional cyclist
400 watts

Maximum power consumption by professional cyclist
(Muscle efficiency=20%)
400 watts/20% = 2,000 watts + 115 watts resting rate

= 2,115 watts power consumption

Metabolism refers to the entire range of many thousands of
interconnected, and subtly regulated, biochemical processes
that happen within a person or living organism that enable it
to keep living, growing and dividing. Metabolic processes can
be divided into:

• **catabolism** – obtaining energy and reducing power from
  nutrients.

• **anabolism** – production of new cell components, usually
  through processes that require energy and reducing power
  obtained from nutrient catabolism.

CELL RESPIRATION

The citric acid (Krebs) cycle releases
energy which is stored as ATP
(adenosine triphosphate) the main
energy currency in the body. Other
metabolic processes that use ATP as
an energy source convert it back into
its precursors. ATP is therefore
continuously recycled in organisms:
the human body, which on average
contains only 250 grams of ATP, turns
over its own body weight equivalent in
ATP each day. This means that each
ATP molecule is recycled 500 to 750
times during a single day.

Digestion

FOOD

proteins   polysaccharides   fats

amino acids   simple sugars   fatty acids
                                            and glycerol

glucose

ATP
NADH

pyruvate

acetyl CoA

CITRIC
ACID
CYCLE

Cell
Membrane

Cytosol

Reducing power
as NADH

Mitochondria

ATP
ATP
ATP

O$_2$

NH$_3$ H$_2$O   CO$_2$

WASTE PRODUCTS

A core set of energy-producing catabolic
pathways occurs within all living organisms in
some form. These pathways transfer the energy
released by breakdown of nutrients into ATP and
other small molecules used for energy.

Schematic map of all human metabolic parthways

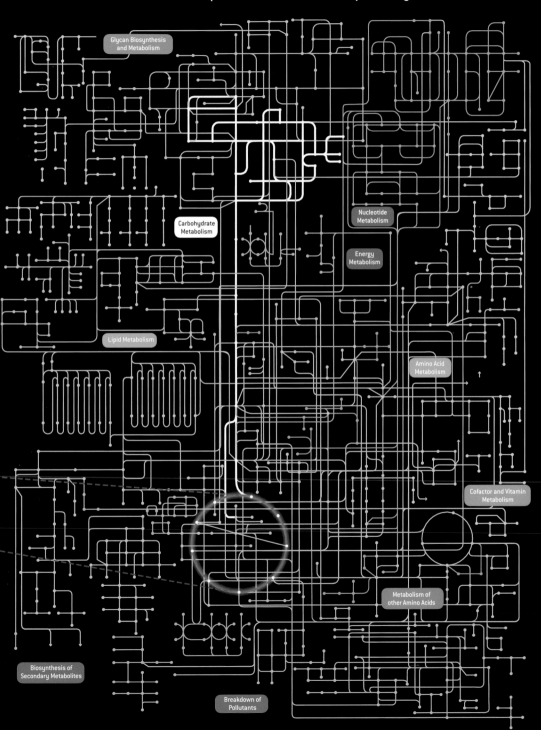

Glycan Biosynthesis
and Metabolism

Carbohydrate
Metabolism

Nucleotide
Metabolism

Energy
Metabolism

Lipid Metabolism

Amino Acid
Metabolism

Cofactor and Vitamin
Metabolism

Metabolism of
other Amino Acids

Biosynthesis of
Secondary Metabolites

Breakdown of
Pollutants

Enzymes

The chemistry of life is based on enzymes – protein molecules that are biological catalysts with three characteristics: they increase the rate of reactions, they act specifically upon only one reactant (called a substrate) to produce products, and most remarkably, they can be regulated ... switched from a state of low activity to high activity and back again.

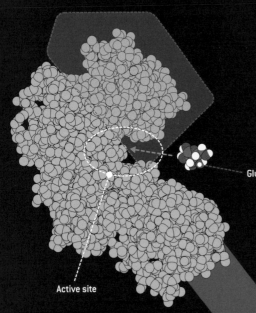

Glucose

Active site

The unique range of up to 3,000 enzymes that each of our cells is genetically programmed to produce, and their systematic regulation, determines the character, shape and function of the cell. If just one enzyme is defective or missing, the outcome can be fatal. The protein molecules that form enzymes vary enormously in molecular weight from 10,000 to 2,000,000 ... the molecular weight of water is 18.

The chemical reactions form the basis of life, and generally move incredibly slowly under normal damp conditions without enzymes. Metabolism is accelerated by enzymes that can speed up chemical reactions by an astonishing ten orders of magnitude or more. For example:

• the spontaneous breaking of the peptide bonds that make up proteins would take 400 years at room temperature

• the break-up of ATP – the energy chip of all living cells – would take about a million years in the absence of enzymes.

CONFORMATIONAL CHANGE

Enzymes change shape as the substrate moves against and then binds with the active site on the enzyme by a process called induced fit, thus forming the enzyme–substrate complex. This is an illustration of the conformational change that occurs when glucose binds to the enzyme hexokinase.

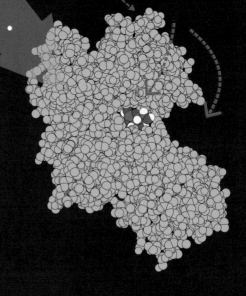

MOLECULES MOVE AT LIGHTNING SPEED

Molecules move extremely quickly due to random thermal motion. A small molecule such as glucose moves around a cell at about 400 kilometres per hour, while large proteins move at over 30 kilometres per hour. At cellular scales this is extremely fast: if a standard human cell were scaled up to the size of Trafalgar Square, large proteins would be moving around at up to 165 million kilometres per hour. In the confines of a cell, molecules can't travel very far without colliding with something. As a result, enzymes can collide with something to react with up to 500,000 times every second.

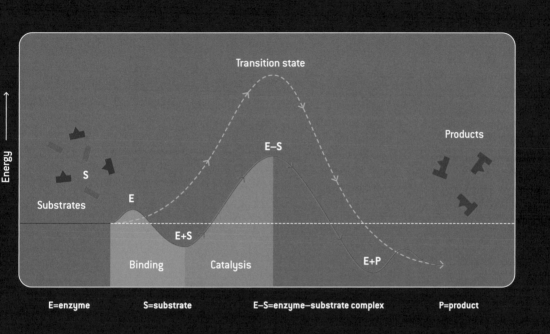

Energy

Transition state

Products

E—S

Substrates

E

E+S

Binding

Catalysis

E+P

E=enzyme   S=substrate   E—S=enzyme—substrate complex   P=product

enzyme—substrate complex

The immune system: biological self-recognition

The immune system has the remarkable ability to distinguish between the body's own cells – recognised as 'self' – and foreign organisms and compounds. When immune defenders encounter markers that say 'non-self', they quickly launch an attack.

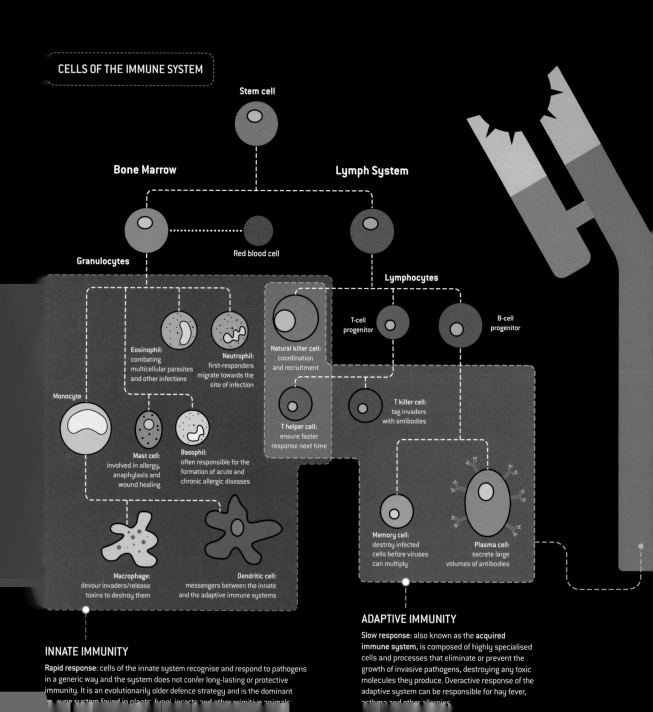

CELLS OF THE IMMUNE SYSTEM

Stem cell

Bone Marrow

Lymph System

Granulocytes

Red blood cell

Lymphocytes

Eosinophil: combating multicellular parasites and other infections

Neutrophil: first-responders migrate towards the site of infection

Natural killer cell: coordination and recruitment

T-cell progenitor

B-cell progenitor

Monocyte

Mast cell: involved in allergy, anaphylaxis and wound healing

Basophil: often responsible for the formation of acute and chronic allergic diseases

T helper cell: ensure faster response next time

T killer cell: tag invaders with antibodies

Macrophage: devour invaders/release toxins to destroy them

Dendritic cell: messengers between the innate and the adaptive immune systems

Memory cell: destroy infected cells before viruses can multiply

Plasma cell: secrete large volumes of antibodies

INNATE IMMUNITY

Rapid response: cells of the innate system recognise and respond to pathogens in a generic way and the system does not confer long-lasting or protective immunity. It is an evolutionarily older defence strategy and is the dominant immune system found in plants, fungi, insects and other primitive animals

ADAPTIVE IMMUNITY

Slow response: also known as the **acquired immune system**, is composed of highly specialised cells and processes that eliminate or prevent the growth of invasive pathogens, destroying any toxic molecules they produce. Overactive response of the adaptive system can be responsible for hay fever, asthma and other allergies

IMMUNOGLOBULINS

also known as antibodies, are glycoprotein molecules produced by plasma cells (white blood cells). They act as a critical part of the immune response by specifically recognising and binding to particular antigens, such as specific proteins on the surface of bacteria or viruses, and aiding in their destruction.

ANTIGENS

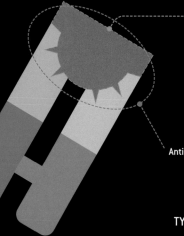

Antigen binding site

Antigens are substances that cause the immune system to produce antibodies. The immune system carries out structural changes to a region of the immunoglobulins, called the **antigen binding site**, that match the antigen so that it can bind to it. Antigens are often free-floating toxins, or specific proteins attached to the membranes of bacteria and viruses. Once attached to the antigen, the antibodies act in various ways: by neutralising and masking dangerous parts of the invading microorganism; by precipitating soluble toxic antigens; or by binding invading cells together.

TYPES OF IMMUNOGLOBULIN

Immunoglobulin G (IgG)

- Typical, Y-shape antibody monomers with two antigen binding sites
- main antibodies in blood: 75% of serum antibodies in humans
- created and released by plasma cells.

Immunoglobulin E (IgE)

- Y-shape monomers
- involved in allergies when the immune system overreacts to an allergen producing IgE that travel to cells that release chemicals, causing an allergic reaction
- main function is defending against parasites.

Immunoglobulin A (IgA)

- dimers (two units), found in mucosa in saliva, the intestines, urinary and respiratory tracts – major sites of potential attack by invading microorganisms
- 3–5 grams secreted into the intestine each day, amounting to up to 15% of the total immunoglobulin produced in the entire body
- an important first line of defence.

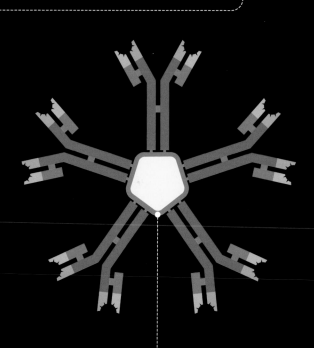

Immunoglobulin M (IgM)

- the first antibodies to respond to infection
- multiple immunoglobulins are covalently linked together, mostly as a pentamer (five units) but sometimes as a hexamer (six units)
- physically the largest antibody in the human circulatory system with a molecular mass of approximately 970,000
- the main antibody responsible for the clotting of red blood cells when someone receives a blood transfusion that is not compatible with their blood type.

Chromosomes: structure and packing

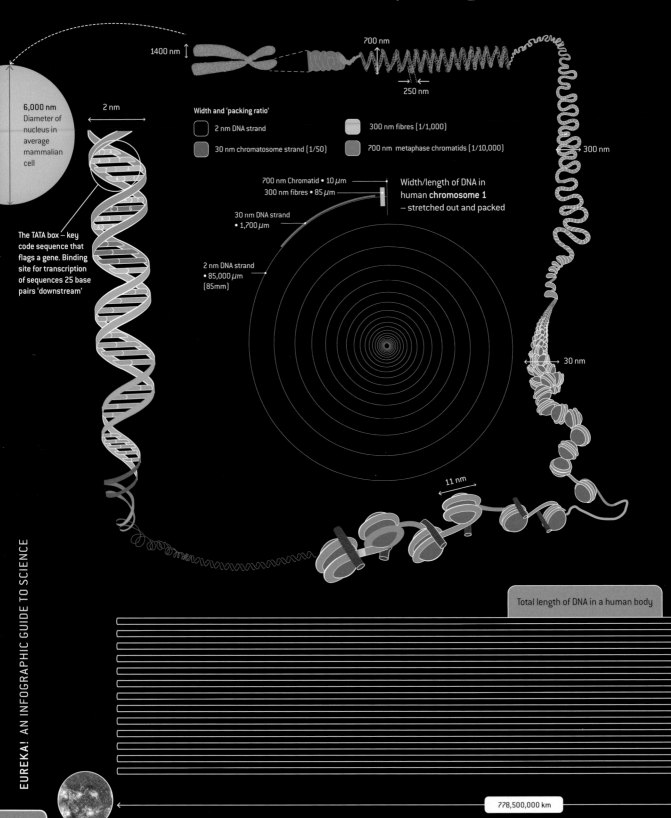

1400 nm

700 nm

250 nm

6,000 nm
Diameter of
nucleus in
average
mammalian
cell

2 nm

Width and 'packing ratio'

2 nm DNA strand

30 nm chromatosome strand (1/50)

300 nm fibres (1/1,000)

700 nm metaphase chromatids (1/10,000)

300 nm

700 nm Chromatid • 10 μm
300 nm fibres • 85 μm

Width/length of DNA in
human **chromosome 1**
– stretched out and packed

30 nm DNA strand
• 1,700 μm

2 nm DNA strand
• 85,000 μm
(85mm)

The TATA box – key
code sequence that
flags a gene. Binding
site for transcription
of sequences 25 base
pairs 'downstream'

30 nm

11 nm

Total length of DNA in a human body

778,500,000 km

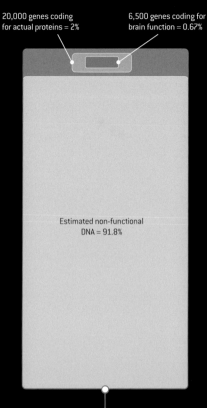

20,000 genes coding for actual proteins = 2%

6,500 genes coding for brain function = 0.67%

Estimated non-functional DNA = 91.8%

Total Human Genome = 3.5 billion base-pairs

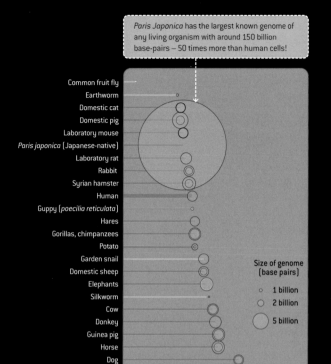

Paris Japonica has the largest known genome of any living organism with around 150 billion base-pairs — 50 times more than human cells!

Common fruit fly
Earthworm
Domestic cat
Domestic pig
Laboratory mouse
Paris japonica (Japanese-native)
Laboratory rat
Rabbit
Syrian hamster
Human
Guppy (*poecilia reticulata*)
Hares
Gorillas, chimpanzees
Potato
Garden snail
Domestic sheep
Elephants
Silkworm
Cow
Donkey
Guinea pig
Horse
Dog
Hedgehog
Goldfish
Kingfisher

Size of genome (base pairs)

○ 1 billion
◯ 2 billion
◯ 5 billion

0   25   50   75   100   125

Number of chromosomes in each cell

TOTAL LENGTH OF DNA IN HUMAN BODY:

Average human cell has 2 metres of chromosomal DNA packed into the nucleus.

Body is estimated to have 10 trillion cells
> total length of stretched-out DNA = 20 billion km
Sun–Jupiter distance (5.2 AU) = 778,500,000 km

Total length of DNA = 137 AU

Equals just over 13 round-trips between Sun and Jupiter

The Genome Enigma

At the heart of the 'Genome enigma' is the puzzle surrounding the great variation in genome size among species and the fact that size does not correlate with complexity. For instance, single-celled organisms such as amoeba have genomes far larger than that of humans. The phenomenon raises questions relating to which bits of an organism's DNA actually code for biochemical and structural complexity, where non-coding DNA comes from and why some organisms have such streamlined genomes while others possess huge quantities of apparently redundant DNA.

Cell reproduction

We grow and repair our bodies through the balletic events that take place as each 'mother' cell duplicates its DNA and divides equally to produce two daughter cells.

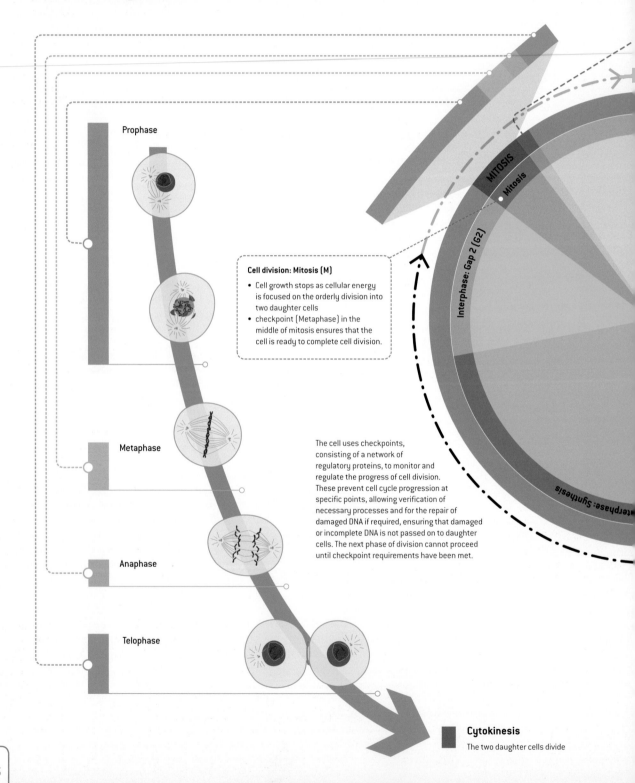

Prophase

Metaphase

Anaphase

Telophase

Cell division: Mitosis (M)

- Cell growth stops as cellular energy is focused on the orderly division into two daughter cells
- checkpoint (Metaphase) in the middle of mitosis ensures that the cell is ready to complete cell division.

The cell uses checkpoints, consisting of a network of regulatory proteins, to monitor and regulate the progress of cell division. These prevent cell cycle progression at specific points, allowing verification of necessary processes and for the repair of damaged DNA if required, ensuring that damaged or incomplete DNA is not passed on to daughter cells. The next phase of division cannot proceed until checkpoint requirements have been met.

MITOSIS

Mitosis

Interphase: Gap 2 (G2)

Interphase: Synthesis

Cytokinesis

The two daughter cells divide

HUMAN CELL LIFESPANS

White blood cells: more than a year
Red blood cells: four months
Skin cells: about two or three weeks
Colon cells: four days

Sperm cells: about three days
Brain cells: an entire lifetime ...
... neurons in the cerebral cortex
are not replaced when they die :(

Cytokinesis

Interphase: Gap 1 (G1)

- cells increase in size
- checkpoint control mechanism
  ensures that everything is ready
  for DNA synthesis.

Interphase: Gap 1 (G1)

Non-dividing state (G0)

INTERPHASE

Gap 0 (G0)

- quiescent/senescent
- A resting phase where the cell
  has left the cycle and has
  stopped dividing.

Interphase: Synthesis (S)

- during this phase the DNA in the nucleus is replicated

CELL SENESCENCE AND DEATH

Multicellular organisms replace worn-out cells through cell division. In some animals, however, division of a cell line will eventually cease. In humans this occurs, on average, after about 50 divisions. The cell is then referred to as senescent. Cells stop dividing because the telomeres, protective end sequences of DNA at each end of a chromosome required for replication, shorten with each copy and are eventually consumed. Research into extending human life often focuses on protecting or rebuilding the telomeres.

CELL CYCLES IN DIFFERENT CELLS AND SPECIES

The speed of cell reproduction varies enormously at different times in the life of an organism, in different kinds of cell tissue, and from species to species. Cleaving cells in a human embryo can divide every 12 hours, while adult liver cells can take up to a year to reproduce. The length of time spent in each part of the cycle can also vary considerably.

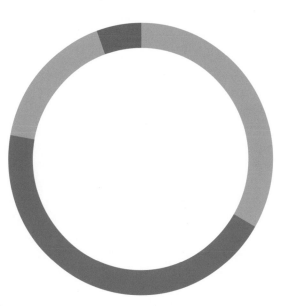

Human cell (20 hours)

Splitting yeast (2 hours)

Budding yeast (2 hours)

Fruit fly embryo (8 minutes)

Sexual reproduction and variation

Mutations are changes in the DNA sequence of a cell's genome – mutation is the ultimate source of variation and without variation there can be no evolution. Mutation is random ... its effects acquire a sense of direction through heritability and environmental fitness.

RECOMBINATION DURING MEIOSIS

During meiosis the genome is divided in to two, with each half producing a gamete that will become a sperm or an egg. During the process maternal and paternal DNA can be exchanged.

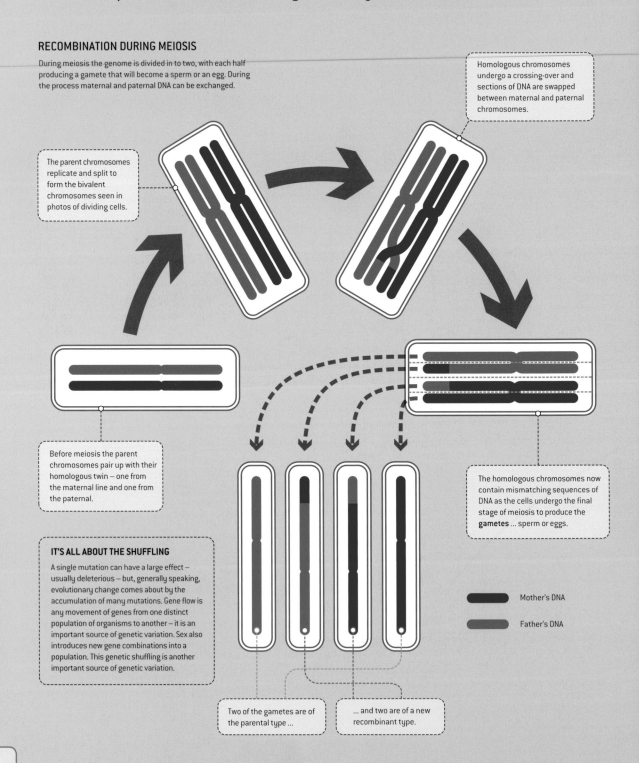

Homologous chromosomes undergo a crossing-over and sections of DNA are swapped between maternal and paternal chromosomes.

The parent chromosomes replicate and split to form the bivalent chromosomes seen in photos of dividing cells.

Before meiosis the parent chromosomes pair up with their homologous twin – one from the maternal line and one from the paternal.

The homologous chromosomes now contain mismatching sequences of DNA as the cells undergo the final stage of meiosis to produce the **gametes** ... sperm or eggs.

IT'S ALL ABOUT THE SHUFFLING

A single mutation can have a large effect – usually deleterious – but, generally speaking, evolutionary change comes about by the accumulation of many mutations. Gene flow is any movement of genes from one distinct population of organisms to another – it is an important source of genetic variation. Sex also introduces new gene combinations into a population. This genetic shuffling is another important source of genetic variation.

Mother's DNA

Father's DNA

Two of the gametes are of the parental type ...

... and two are of a new recombinant type.

EVOLUTION THROUGH VARIATION AND NATURAL SELECTION
The combination of small-scale mutation, at the level of genes and DNA base sequences, and chromosomal recombination introduces variation from generation to generation. In this case the environment favours offspring that are rounder and darker. Pale squares must die!

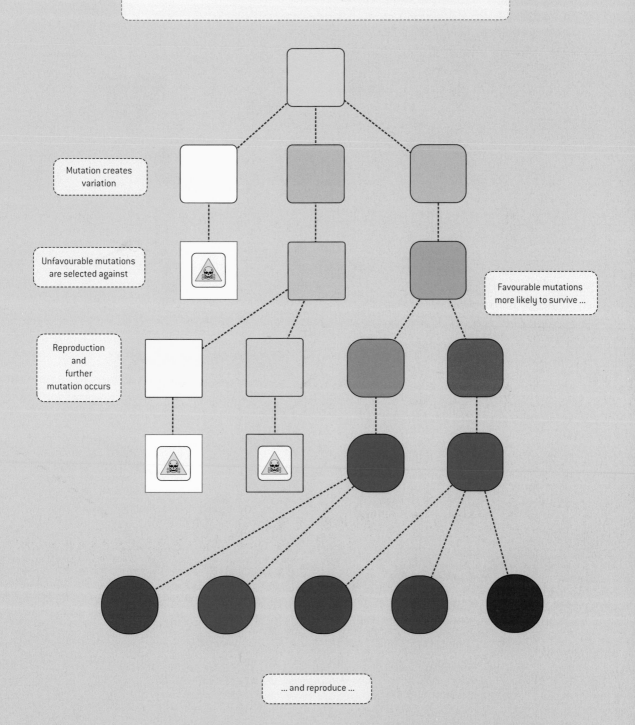

Mutation creates variation

Unfavourable mutations are selected against

Favourable mutations more likely to survive ...

Reproduction and further mutation occurs

... and reproduce ...

Bacteria

Bacteria were here at the dawn of life and are still going strong. They biochemically adapt –
hence antibiotic resistance – at an astounding rate, and yet they have remained structurally
unchanged for billions of years.

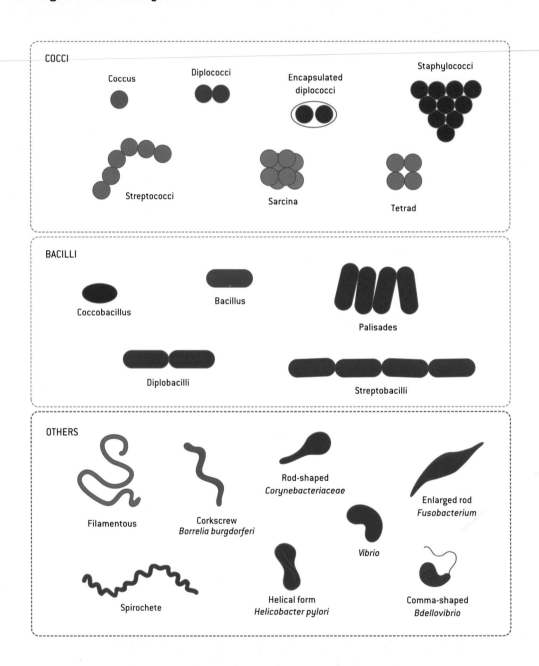

Although there are many thousands of species of bacteria, they all share three basic shapes. Some are shaped like
sticks or rods, called **bacilli**. Others are shaped as little balls and called **cocci**. Still others have a helical or spiral
shape, called **spirochetes**. Bacteria can exist as an individual bacterium while others group together to form pairs,
chains and other connected structures.

THE HUMAN MICROBIOTA

Estimates for the total number of bacterial cells found in association with the human body have varied between 10 and 1.5 bacteria for each and every human cell. The total number of bacterial genes associated with the human microbiota could exceed the total number of human genes by a factor of 80 to 1. Conservative estimates suggest that an average 70 kg human being is composed of about 30 trillion human cells … and 40 trillion bacterial cells.

5,000,000,000,000,000,000,000,000,000,000

Estimated number of bacteria on the planet – that's over 70 million bacteria for every star in the Universe

MAP OF HUMAN SKIN BACTERIA

The human body provides a rich and varied environment for bacteria. Different parts of the body host very different communities of prokaryotes. Most of the microbes appear not to be too harmful, and many assist in maintaining processes necessary for a healthy body.

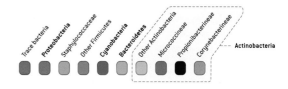

Trace bacteria · **Proteobacteria** · Staphylococcaceae · Other Firmicutes · **Cyanobacteria** · **Bacteroidetes** · Other Actinobacteria · Micrococcineae · Propionibacterineae · Corynebacterineae — Actinobacteria

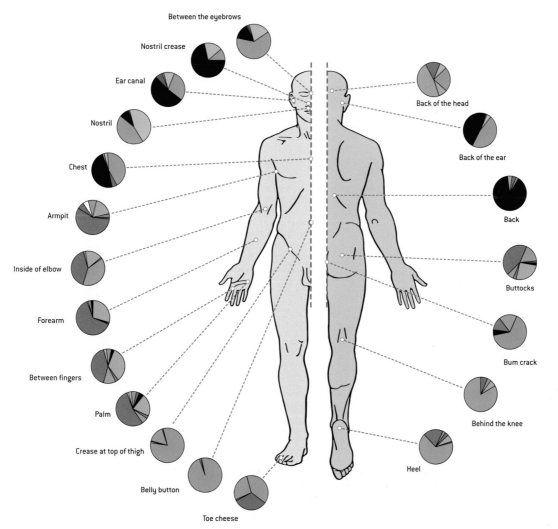

Between the eyebrows · Nostril crease · Ear canal · Nostril · Chest · Armpit · Inside of elbow · Forearm · Between fingers · Palm · Crease at top of thigh · Belly button · Toe cheese · Back of the head · Back of the ear · Back · Buttocks · Bum crack · Behind the knee · Heel

Single-cell life

The earliest eukaryotic life was single-celled, but at a certain point multicellularity evolved and the extraordinary, diverse and beautiful macroscopic realm of life developed. One theory is that colonial protists evolved in to multicellular animals such as sponges and comb jellies, the most primitive true animals.

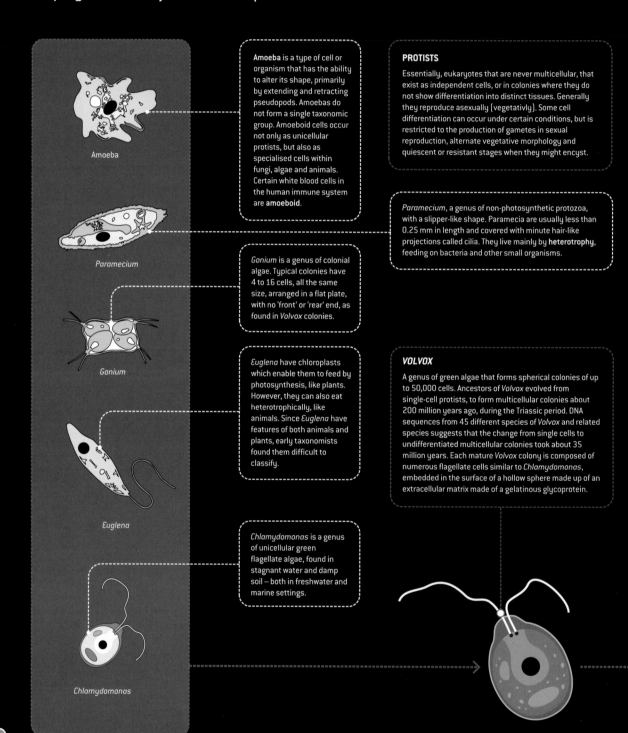

Amoeba

Paramecium

Gonium

Euglena

Chlamydomonas

Amoeba is a type of cell or organism that has the ability to alter its shape, primarily by extending and retracting pseudopods. Amoebas do not form a single taxonomic group. Amoeboid cells occur not only as unicellular protists, but also as specialised cells within fungi, algae and animals. Certain white blood cells in the human immune system are **amoeboid**.

Gonium is a genus of colonial algae. Typical colonies have 4 to 16 cells, all the same size, arranged in a flat plate, with no 'front' or 'rear' end, as found in *Volvox* colonies.

Euglena have chloroplasts which enable them to feed by photosynthesis, like plants. However, they can also eat heterotrophically, like animals. Since *Euglena* have features of both animals and plants, early taxonomists found them difficult to classify.

Chlamydomonas is a genus of unicellular green flagellate algae, found in stagnant water and damp soil – both in freshwater and marine settings.

PROTISTS

Essentially, eukaryotes that are never multicellular, that exist as independent cells, or in colonies where they do not show differentiation into distinct tissues. Generally they reproduce asexually (vegetativly). Some cell differentiation can occur under certain conditions, but is restricted to the production of gametes in sexual reproduction, alternate vegetative morphology and quiescent or resistant stages when they might encyst.

Paramecium, a genus of non-photosynthetic protozoa, with a slipper-like shape. Paramecia are usually less than 0.25 mm in length and covered with minute hair-like projections called cilia. They live mainly by **heterotrophy**, feeding on bacteria and other small organisms.

VOLVOX

A genus of green algae that forms spherical colonies of up to 50,000 cells. Ancestors of *Volvox* evolved from single-cell protists, to form multicellular colonies about 200 million years ago, during the Triassic period. DNA sequences from 45 different species of *Volvox* and related species suggests that the change from single cells to undifferentiated multicellular colonies took about 35 million years. Each mature *Volvox* colony is composed of numerous flagellate cells similar to *Chlamydomonas*, embedded in the surface of a hollow sphere made up of an extracellular matrix made of a gelatinous glycoprotein.

SPONGE

COMB JELLY

In contrast to unicellular colonies, even the simplest multicellular organisms have cells that depend on each other to survive. Most multicellular organisms have a unicellular life-cycle stage. Gametes, eggs and sperm, for example, are reproductive unicells for multicellular organisms. Multicellularity appears to have evolved independently many times in the history of life. The most primitive true **metazoans** are the **sponges** and **comb jellies**.

The individual *Volvox* cells are connected to each other by thin strands of cytoplasm that enable the whole colony to swim in a coordinated fashion. The individual alga also have small red eye spots. The colonies even have a front and rear end: the 'front end' has more developed eyespots which enables the colony to swim towards the light. This differentiation of cells make *Volvox* quite unique. It is a colony that comes really close to being a **multicellular organism**.

Volvox colonies can grow to over 2 mm in diameter, so they can be easily seen with the naked eye

It can often be hard to distinguish colonial protists from true multicellular organisms, because the two concepts are not wholly distinct – colonial protists have been dubbed **pluricellular** rather than multicellular. They sit between the two great divisions of eukaryotic life.

Most *Volvox* colonies have spheres inside. These are 'daughter' colonies which grow asexually from cells around the middle of the colony. These cells enlarge and undergo a series of cell divisions until they form a small sphere. The daughter colonies have to turn themselves before the parent colony releases them into the surrounding water.

Precambrian life: the Ediacaran biota

Up until the 1950s there was an almost universally held belief that complex life originated after the start of the Cambrian era, some 540 million years ago. More recently, the fossil record has revealed a long history of life displaying a vast range of morphological characteristics, with body sizes ranging from millimetres to metres.

The Ediacaran biota are distinguished by the fact that none of these organisms seemed to go on to evolve into any of the known living phyla we see today ... or in later fossil records. Some Ediacaran organisms have more complex details preserved, such as *Vernanimalcula*, which has led to them being controversially interpreted as possible early forms of living organisms, excluding them from some definitions of the Ediacaran biota.

Vernanimalcula

Ratio of carbon-12 to carbon-13: living organisms preferentially take up carbon-12, so leave a signature ratio of the two isotopes lower in carbon-13. This is a good indicator of primary production and organic burial ... the scale and vigour of the planet's biosphere.

The first Ediacaran fossils, discovered in 1868, were the disc-shaped *Aspidella terranovica*. They have been dated from 610 to 555 million years ago, with disputed representatives dating back 770 million years.

Microfossils from 632 million years ago – only three million years after the end of the Marinoan Glaciation – could be embryonic 'resting stages' in the lifecycle of the earliest-known muliticellular animals. There is some dispute over this interpretation.

?

'Embryos' from the Doushantuo Formation in China.

MARINOAN GLACIATION (lasted 9 million years)

EDIACARAN

NEOPROTEROZOIC/PRECAMBRIAN

640   630   620   610   600

Millions of years before the present

Tribrachidium

PRESERVATION
The preservation of these fossils is a subject of great interest and debate. As soft-bodied organisms, they would normally not fossilise and, unlike later soft-bodied fossil biota found in the Burgess Shales or Solnhofen Limestone in Germany, the Ediacaran fossils are not found in very specific environments subject to unusual local conditions: they were a global phenomenon.

Dickinsonia

Kimberella

Spriggina

First bilaterian?

GASKIERS GLACIATION (lasted 4 million years)

BAYKONUR GLACIATION (4 million years)

Fractofusus

Charnia

50 genera

MAIN SERIES OF EDIACARAN FOSSILS

Number of genera in the fossil record. The beginning of the Cambrian Explosion.

Aspidella

EDIACARAN

CAMBRIAN

PALEOZOIC

580   570   560   550   540

Bilaterians

Bilaterians are the animals with bilateral symmetry — they have a front end and a back end, and an upside and downside, and therefore a left and a right. Once you have a front end, you have a head, and if you have a head you have a brain.

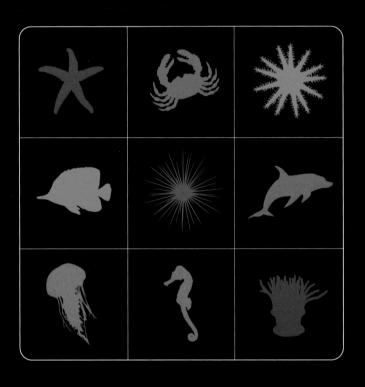

BODY PLANS AND SYMMETRY

The bilaterians are a major group of animals, including the majority of living phyla, but not sponges, jellyfish, anenomes or sea urchins. For the most part, bilateral embryos are triploblastic, having three original cell layers: **endoderm**, **mesoderm**, and **ectoderm**. Nearly all are bilaterally symmetrical, or approximately so – the most notable exception are starfish, which achieve near-radial symmetry as adults, but are bilaterally symmetrical as larvae.

**BASIC BODY PLANS
(CROSS SECTION)**

Increasing complexity

**Diploblastic
Acoelomate
(jellyfish)**
- gut
- endoderm
- ectoderm

**Triploblastic
Acoelomate
(flatworms)**
- gut
- endoderm
- mesoderm
- ectoderm

**Triploblastic
Coelomate
(starfish, molluscs,
vertebrates)**
- gut
- endoderm
- coelom
- mesoderm
  (split in two by coelom)
- ectoderm

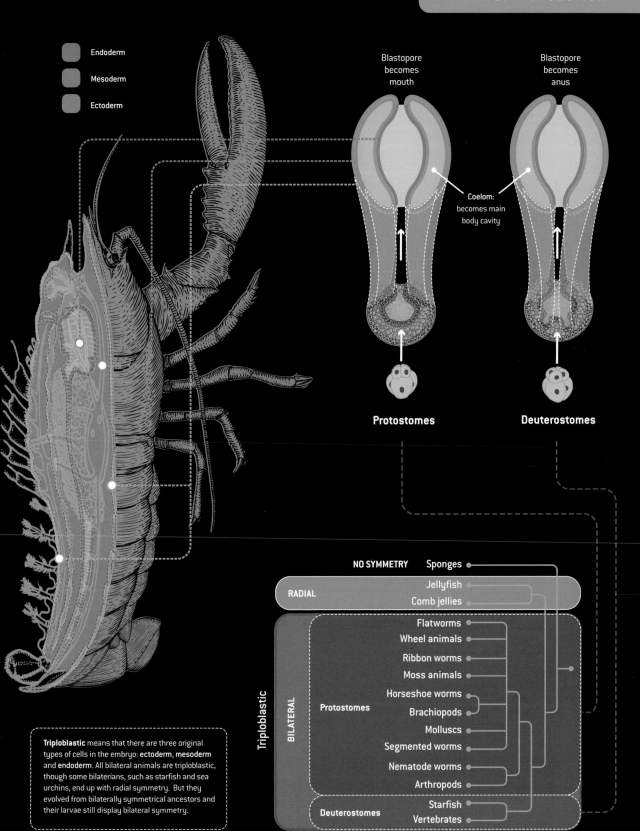

Endoderm

Mesoderm

Ectoderm

Blastopore becomes mouth

Blastopore becomes anus

Coelom: becomes main body cavity

Protostomes

Deuterostomes

NO SYMMETRY — Sponges

RADIAL — Jellyfish / Comb jellies

Triploblastic

BILATERAL

Protostomes
- Flatworms
- Wheel animals
- Ribbon worms
- Moss animals
- Horseshoe worms
- Brachiopods
- Molluscs
- Segmented worms
- Nematode worms
- Arthropods

Deuterostomes
- Starfish
- Vertebrates

Triploblastic means that there are three original types of cells in the embryo: **ectoderm**, **mesoderm** and **endoderm**. All bilateral animals are triploblastic, though some bilaterians, such as starfish and sea urchins, end up with radial symmetry. But they evolved from bilaterally symmetrical ancestors and their larvae still display bilateral symmetry.

Mass extinction events and evolution

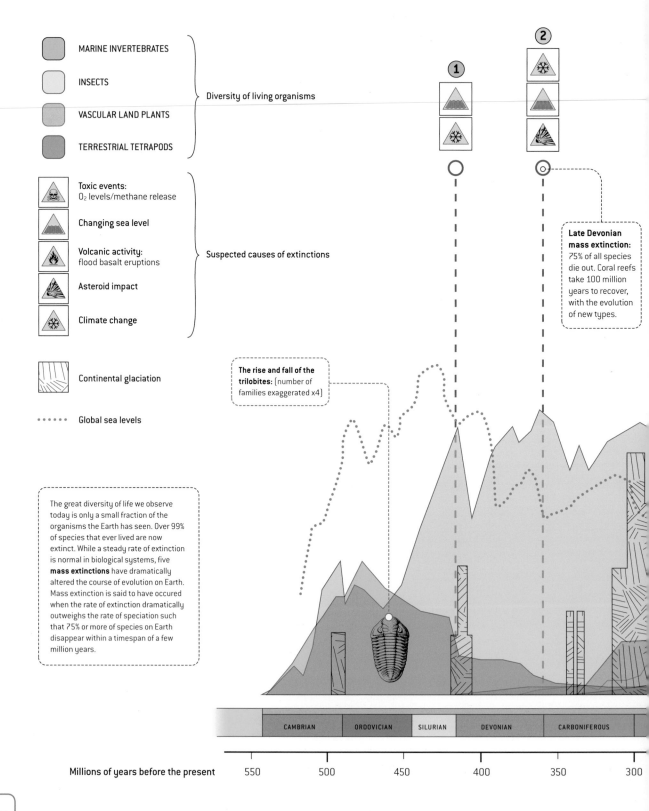

MARINE INVERTEBRATES

INSECTS

VASCULAR LAND PLANTS

TERRESTRIAL TETRAPODS

Diversity of living organisms

Toxic events:
O₂ levels/methane release

Changing sea level

Volcanic activity:
flood basalt eruptions

Asteroid impact

Climate change

Suspected causes of extinctions

Continental glaciation

Global sea levels

The rise and fall of the trilobites: (number of families exaggerated x4)

① ②

Late Devonian mass extinction: 75% of all species die out. Coral reefs take 100 million years to recover, with the evolution of new types.

The great diversity of life we observe today is only a small fraction of the organisms the Earth has seen. Over 99% of species that ever lived are now extinct. While a steady rate of extinction is normal in biological systems, five **mass extinctions** have dramatically altered the course of evolution on Earth. Mass extinction is said to have occured when the rate of extinction dramatically outweighs the rate of speciation such that 75% or more of species on Earth disappear within a timespan of a few million years.

| CAMBRIAN | ORDOVICIAN | SILURIAN | DEVONIAN | CARBONIFEROUS |

Millions of years before the present   550   500   450   400   350   300

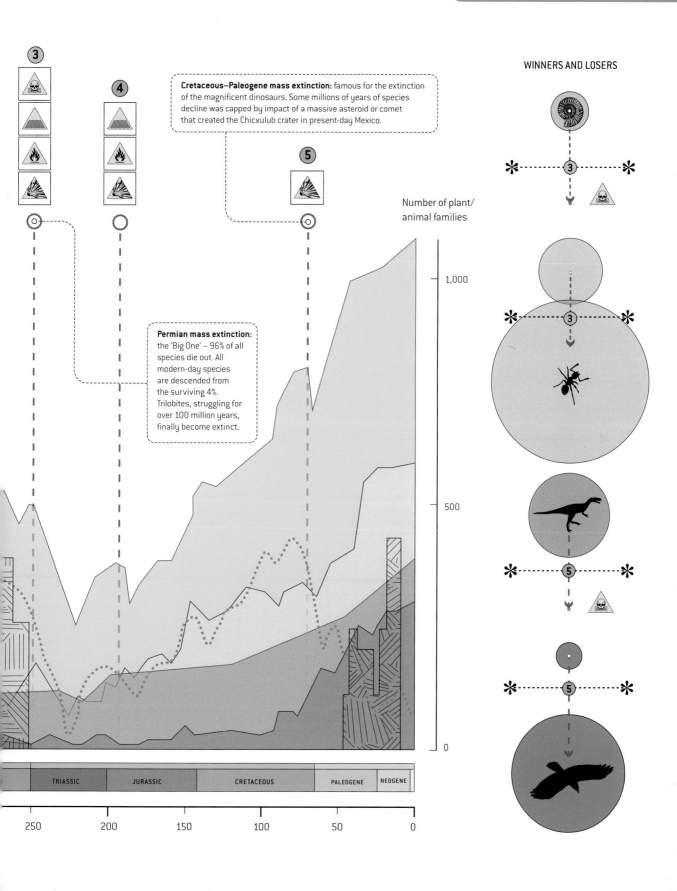

WINNERS AND LOSERS

Cretaceous–Paleogene mass extinction: famous for the extinction of the magnificent dinosaurs. Some millions of years of species decline was capped by impact of a massive asteroid or comet that created the Chicxulub crater in present-day Mexico.

Permian mass extinction: the 'Big One' – 96% of all species die out. All modern-day species are descended from the surviving 4%. Trilobites, struggling for over 100 million years, finally become extinct.

Number of plant/ animal families

1,000

500

0

| TRIASSIC | JURASSIC | CRETACEOUS | PALEOGENE | NEOGENE |

250   200   150   100   50   0

The Cambrian explosion

The Cambrian explosion was a relatively short evolutionary period during which almost all of the major animal phyla appeared in the fossil record. Prior to the Cambrian explosion, life was mostly composed of simple, single cells occasionally organized into colonies.

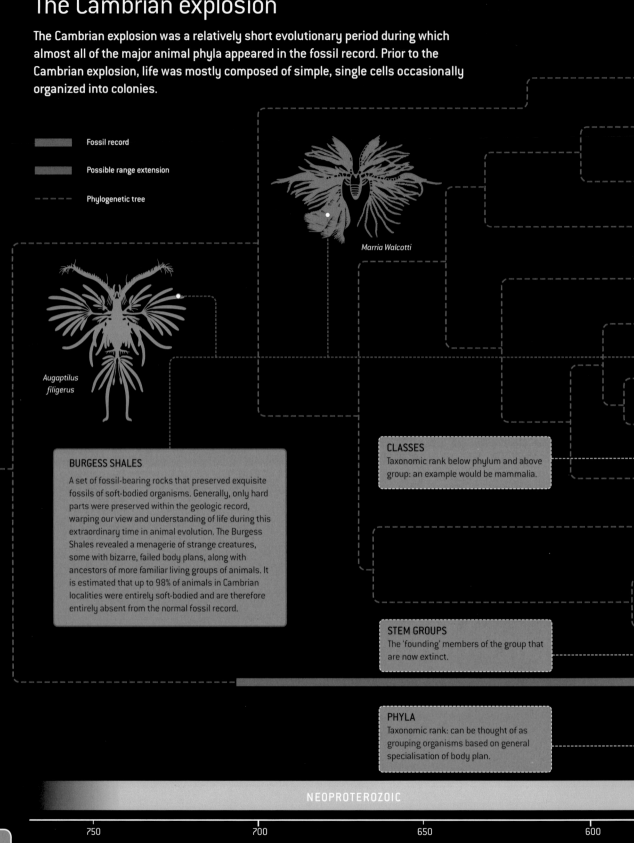

Fossil record

Possible range extension

Phylogenetic tree

Marria Walcotti

Augaptilus filigerus

BURGESS SHALES

A set of fossil-bearing rocks that preserved exquisite fossils of soft-bodied organisms. Generally, only hard parts were preserved within the geologic record, warping our view and understanding of life during this extraordinary time in animal evolution. The Burgess Shales revealed a menagerie of strange creatures, some with bizarre, failed body plans, along with ancestors of more familiar living groups of animals. It is estimated that up to 98% of animals in Cambrian localities were entirely soft-bodied and are therefore entirely absent from the normal fossil record.

CLASSES
Taxonomic rank below phylum and above group: an example would be mammalia.

STEM GROUPS
The 'founding' members of the group that are now extinct.

PHYLA
Taxonomic rank: can be thought of as grouping organisms based on general specialisation of body plan.

NEOPROTEROZOIC

750   700   650   600

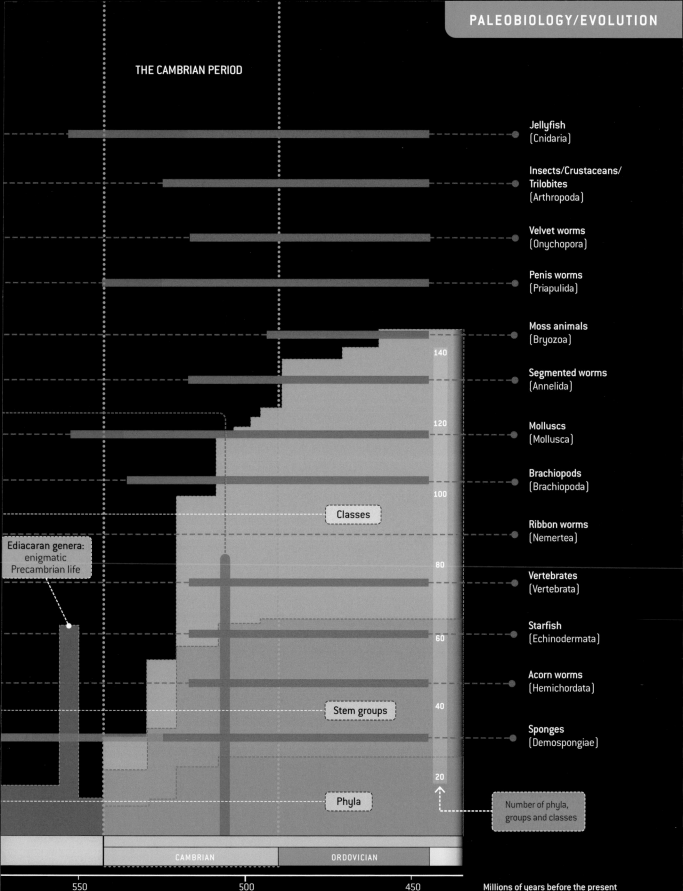

THE CAMBRIAN PERIOD

Jellyfish
(Cnidaria)

Insects/Crustaceans/
Trilobites
(Arthropoda)

Velvet worms
(Onychopora)

Penis worms
(Priapulida)

Moss animals
(Bryozoa)

140

Segmented worms
(Annelida)

120

Molluscs
(Mollusca)

Brachiopods
(Brachiopoda)

100

Classes

Ribbon worms
(Nemertea)

Ediacaran genera:
enigmatic
Precambrian life

80

Vertebrates
(Vertebrata)

Starfish
(Echinodermata)

60

Acorn worms
(Hemichordata)

40

Stem groups

Sponges
(Demospongiae)

20

Phyla

Number of phyla,
groups and classes

CAMBRIAN   ORDOVICIAN

550   500   450   Millions of years before the present

The chordates

The evolutionary bifurcation that led to the chordates was a crucial fork in the road of life. The appearance of a dorsal nerve cord and a rod of cartilage eventually led to the central nervous system and the emergence of the vertebrate brain, the organ that rules the planet.

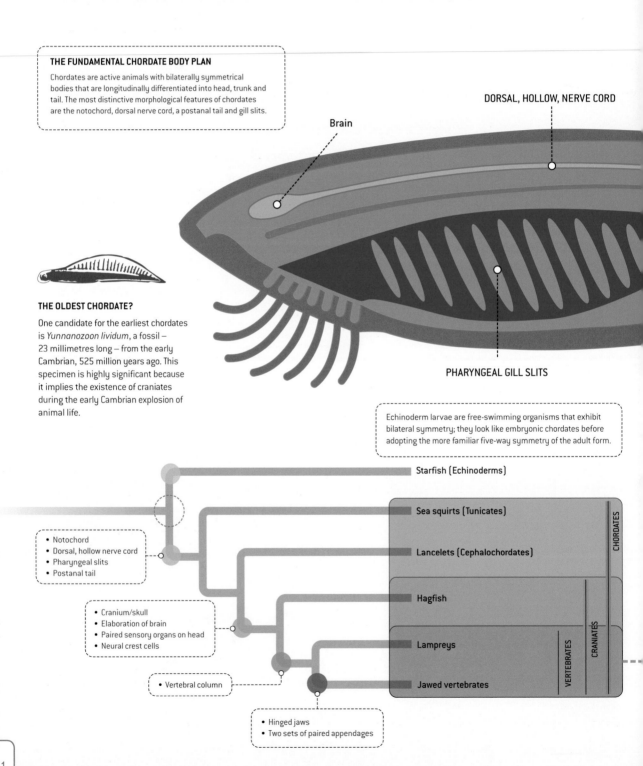

THE FUNDAMENTAL CHORDATE BODY PLAN

Chordates are active animals with bilaterally symmetrical bodies that are longitudinally differentiated into head, trunk and tail. The most distinctive morphological features of chordates are the notochord, dorsal nerve cord, a postanal tail and gill slits.

Brain

DORSAL, HOLLOW, NERVE CORD

THE OLDEST CHORDATE?

One candidate for the earliest chordates is *Yunnanozoon lividum*, a fossil – 23 millimetres long – from the early Cambrian, 525 million years ago. This specimen is highly significant because it implies the existence of craniates during the early Cambrian explosion of animal life.

PHARYNGEAL GILL SLITS

Echinoderm larvae are free-swimming organisms that exhibit bilateral symmetry; they look like embryonic chordates before adopting the more familiar five-way symmetry of the adult form.

Starfish (Echinoderms)

Sea squirts (Tunicates)

- Notochord
- Dorsal, hollow nerve cord
- Pharyngeal slits
- Postanal tail

Lancelets (Cephalochordates)

Hagfish

- Cranium/skull
- Elaboration of brain
- Paired sensory organs on head
- Neural crest cells

Lampreys

Jawed vertebrates

- Vertebral column

- Hinged jaws
- Two sets of paired appendages

CHORDATES

CRANIATES

VERTEBRATES

An advanced nervous system and an internal musculoskeletal system are the key vertebrate innovations. Animal intelligence of a different order was released into the biosphere. While fish don't seem very intelligent compared to apes ... they are when compared to flatworms! The one exception in the invertebrate world is the octopus, which may be equally smart or smarter than quite a few vertebrates. That intelligence, allied with the ability to move with strength, speed and agility, gives vertebrates such as reptiles and birds a huge advantage in the natural world.

LIFE-SIZE

Paedophryne amauensis

A species of frog discovered in Papua New Guinea and formally described in 2012. At 7.7 millimetres long, it is considered to be the world's smallest known vertebrate.

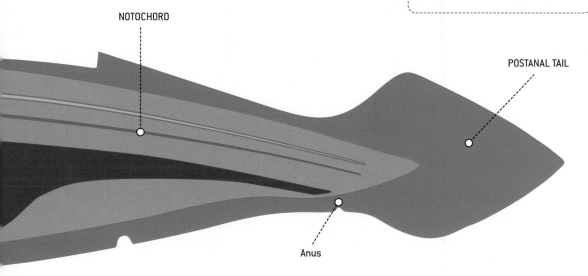

NOTOCHORD

POSTANAL TAIL

Anus

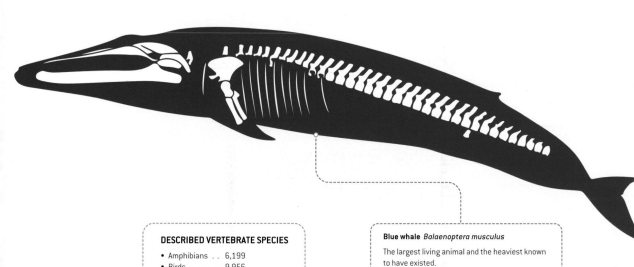

DESCRIBED VERTEBRATE SPECIES

- Amphibians . . 6,199
- Birds 9,956
- Fish 30,000
- Mammals . . . 5,416
- Reptiles 8,240

TOTAL 59,811

Blue whale *Balaenoptera musculus*

The largest living animal and the heaviest known to have existed.
- **length:** up to 30 metres
- **weight:** 73,000–136,000 kilograms
- **lifespan:** approximately 80 years

Fish

During the Devonian period, jawed fish diversified rapidly, with all the modern groups emerging and thriving. With over 30,000 known living species, fish account for more than half of all vertebrate species.

PLACODERMS

An extinct class of armoured prehistoric fish, the largest species measuring up to 10 metres in length and weighing up to 3,600 kilograms. They were hypercarnivorous apex predators and dominated the Devonian period, but, being slower and less agile than evolving sharks and ray-finned fish, they succumbed to competition and a series of environmental shocks.

SPINY SHARKS

Also known as acanthodians, this class of extinct fish has a misleading name – they weren't sharks, but they did share features present in both bony fish and cartilaginous fish. In form they somewhat resembled sharks, but their skin was covered with tiny **rhomboid platelets**.

THE AGE OF THE FISHES

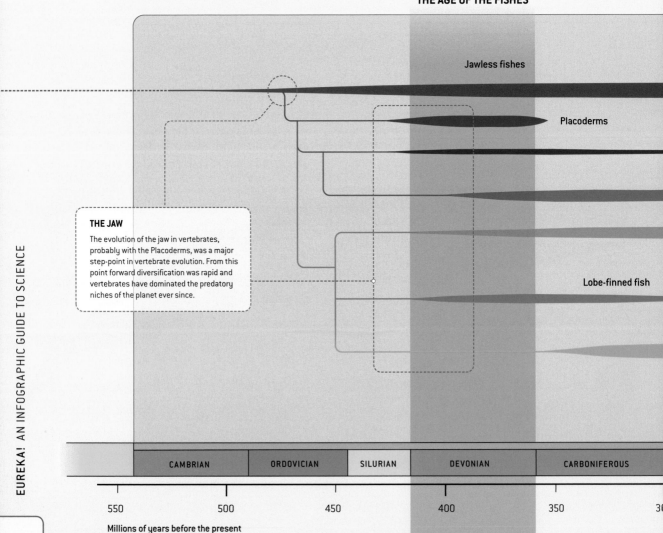

Jawless fishes

Placoderms

Lobe-finned fish

THE JAW

The evolution of the jaw in vertebrates, probably with the Placoderms, was a major step-point in vertebrate evolution. From this point forward diversification was rapid and vertebrates have dominated the predatory niches of the planet ever since.

CAMBRIAN	ORDOVICIAN	SILURIAN	DEVONIAN	CARBONIFEROUS

550   500   450   400   350   3

Millions of years before the present

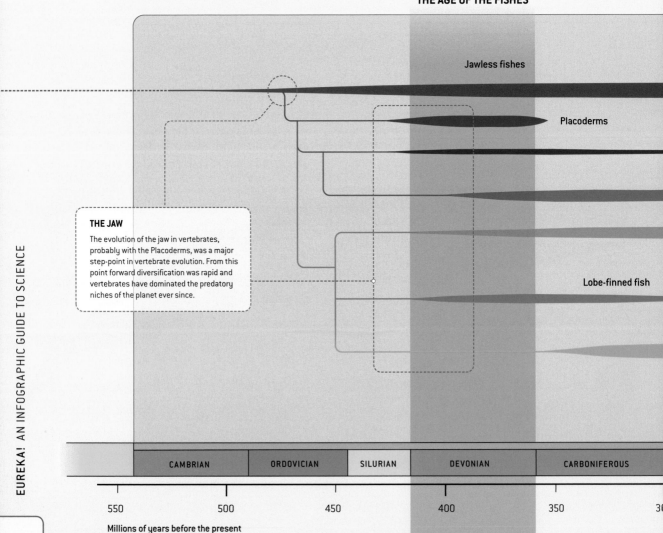

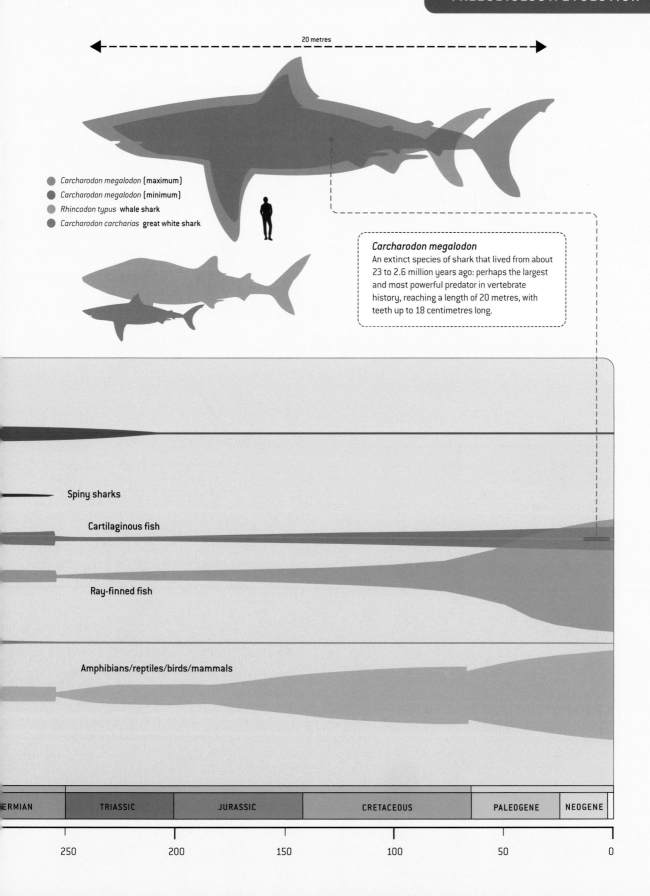

20 metres

● *Carcharodon megalodon* (maximum)
● *Carcharodon megalodon* (minimum)
● *Rhincodon typus* whale shark
● *Carcharodon carcharias* great white shark

Carcharodon megalodon
An extinct species of shark that lived from about 23 to 2.6 million years ago: perhaps the largest and most powerful predator in vertebrate history, reaching a length of 20 metres, with teeth up to 18 centimetres long.

Spiny sharks

Cartilaginous fish

Ray-finned fish

Amphibians/reptiles/birds/mammals

| ERMIAN | TRIASSIC | JURASSIC | CRETACEOUS | PALEOGENE | NEOGENE |

250   200   150   100   50   0

Marine habitats

The world's five oceans comprise 97 per cent of the total water on Earth with a combined volume of over 1,300,000,000 km3. Standing in a column three kilometres wide, the oceans' waters would reach the Sun.

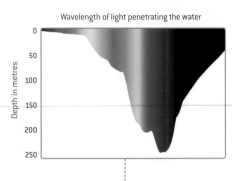

Wavelength of light penetrating the water

Depth in metres

0
50
100
150
200
250

Most marine life is found in coastal habitats, even though the shelf area occupies only seven percent of the total ocean area.

Continental Shelf

Biomass in
water column

200 m

1,000 m

Depth below sea level
in metres

Continental slope

Continental rise

4,000 m

Mount Everest
8,848 m

MARINE HABITATS

- Littoral zone
- Intertidal zone
- Estuaries
- Kelp forests
- Coral reefs
- Ocean banks
- Continental shelf
- Neritic zone
- Straits
- Pelagic zone
- Oceanic zone
- Seamounts
- Hydrothermal vents
- Cold seeps
- Demersal zone
- Benthic zone

10,000 m

EPIPELAGIC

The pelagic zone includes those waters further from the land ... essentially, the open ocean. The epipelagic zone is the illuminated zone at the surface of the sea where enough light is available for photosynthesis. Pelagic photosynthesis accounts for almost 45% of the Earth's carbon fixation, despite the fact that the total living biomass in the ocean is only about one two-hundredth of that in the land plants.

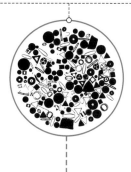

Mauna Kea
>10,000 m

Water temperature (°C)

0   5   10   15   20

Abyssal plain
4,280 m average depth of Pacific
3,926 m average depth of Atlantic

Submarine ridge

Oceanic trench

BATHYPELAGIC FISH

From 1,000–4,000 m the darkness is complete, the pressure is crushing, and temperatures, nutrients and dissolved oxygen levels are all low. Fishlife in this zone are sedentary, adapted to using minimum energy in a habitat with very little food or available energy – despite their ferocious appearance, these beasts of the deep are mostly miniature fish with very weak muscles. Their bodies are elongated with weak, watery muscles and skeletal structures and are slimy, without scales. They often have extensible, hinged jaws with recurved teeth; the eyes are small and may not function.

Gills versus lungs

The larger and more active you are as an animal, the greater the challenges of getting enough oxygen. Land and sea present very different bioengineering challenges.

DIFFUSION RATE

Oxygen diffuses in air at a rate 10,000 times greater than in water.

PROBLEMS ADJUSTING TO LIFE ON LAND

- no buoyancy; outside of a liquid environment more support is necessary for mobility.
- desiccation; danger of drying out
- extreme variation in temperature
- lack of nutrients in air
- refraction of light is different so eyes need to adjust.

VISCOSITY

Water is 100 times more viscous than air.

1 litre of fresh air contains 210 cm^3 of oxygen

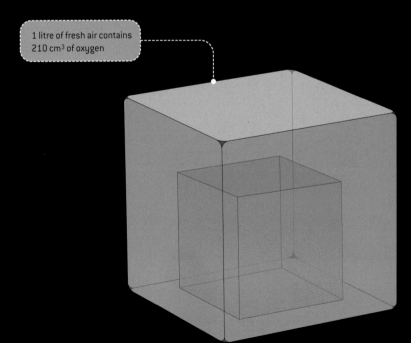

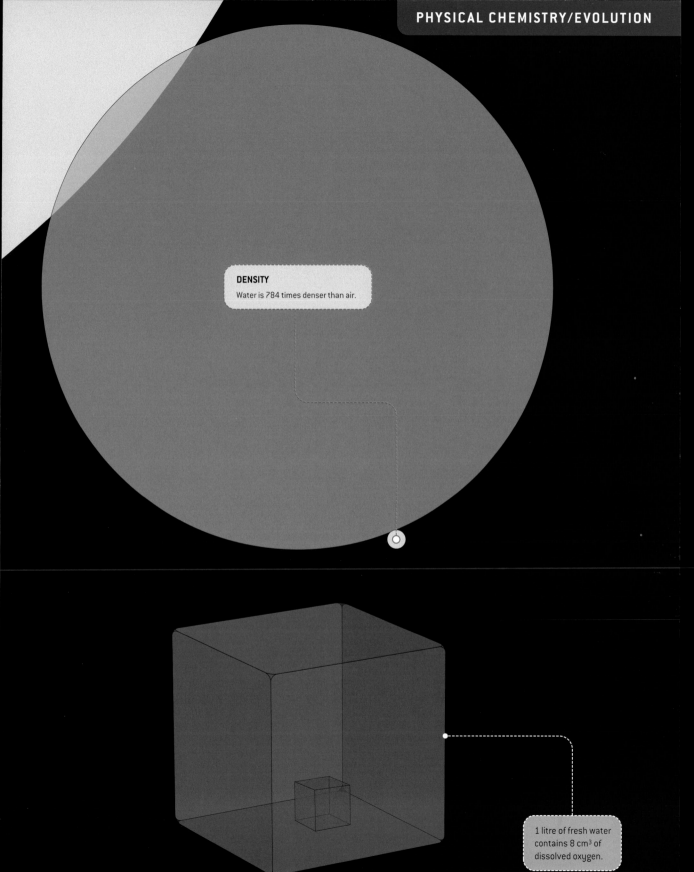

DENSITY
Water is 784 times denser than air.

1 litre of fresh water
contains 8 cm³ of
dissolved oxygen.

Invasion of the land

In the million years between 450 and 350 million years ago, the oceans and lakes gave up their living as first plants, then bugs and finally vertebrates dragged themselves from their watery homes to conquer the vast empty continents.

Ankle-high shrubs

A species of millipede called *Pneumodesmus newmani* that lived 428 million years ago is the oldest known animal to have lived on land.

395 million years ago

The first lichens, stoneworts. Earliest harvestmen, mites, woodlice and ammonoids. The first known tetrapod tracks on land.

450 million years ago

Fossilised spores, suggesting the first land plants, are found in the Ordovician period.

PLANTS AND INVERTEBRATES

434 million years ago

The first primitive plants colonise the land, having evolved from green algae living along the margins of the land. They are accompanied by fungi, which may have assisted the move away from water through symbiotic relationships. The evolution of the first land plants was a major event in the history of the Earth. It opened the way for the irresistible evolution of terrestrial animal life. It was the land plants that changed the biosphere for ever, changing oxygen levels in the atmosphere, the acidity of rivers and lakes, the soil structure and the character of continental erosion.

Panderichthys (380 mya)
Has a large tetrapod-like flattened head, narrow at the snout and wide at the back.

Eusthenopteron (385 mya)
Lobe-finned fish tetrapod progenitor: earliest known fossilised evidence of bone marrow – may represent the origin of bone marrow in tetrapods.

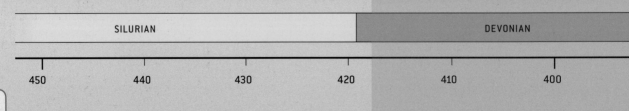

SILURIAN	DEVONIAN

450   440   430   420   410   400

Earliest known tree

First roots, seeds and leaves

First forests

winged insects

bristletails

springtails

harvestmen

mites

centipedes

millipedes

By the early Carboniferous period the Earth many aspects of the biosphere would be recognisable today. Insects roamed the land and had taken to the skies ... sharks swam the seas as top predators, and vegetation covered the land, with seed-bearing and woody, tree-like plants soon to flourish. Four-limbed vertebrates gradually gain adaptations that help them adopt a truly terrestrial life.

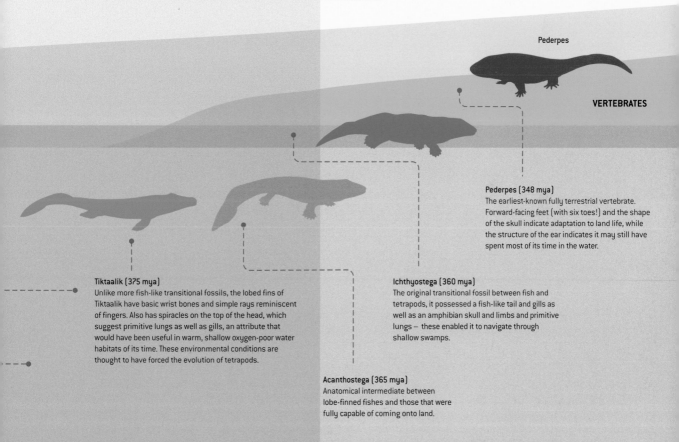

Pederpes

VERTEBRATES

Pederpes (348 mya)
The earliest-known fully terrestrial vertebrate. Forward-facing feet (with six toes!) and the shape of the skull indicate adaptation to land life, while the structure of the ear indicates it may still have spent most of its time in the water.

Tiktaalik (375 mya)
Unlike more fish-like transitional fossils, the lobed fins of Tiktaalik have basic wrist bones and simple rays reminiscent of fingers. Also has spiracles on the top of the head, which suggest primitive lungs as well as gills, an attribute that would have been useful in warm, shallow oxygen-poor water habitats of its time. These environmental conditions are thought to have forced the evolution of tetrapods.

Ichthyostega (360 mya)
The original transitional fossil between fish and tetrapods, it possessed a fish-like tail and gills as well as an amphibian skull and limbs and primitive lungs — these enabled it to navigate through shallow swamps.

Acanthostega (365 mya)
Anatomical intermediate between lobe-finned fishes and those that were fully capable of coming onto land.

DEVONIAN	CARBONIFEROUS

380   370   360   350   340   330

Millions of years before the present

Carboniferous period

During this period (360–300 million years ago), amphibians were the dominant land vertebrates — one branch would eventually evolve into reptiles, the first fully terrestrial vertebrates — and vast swathes of forest covered the land.

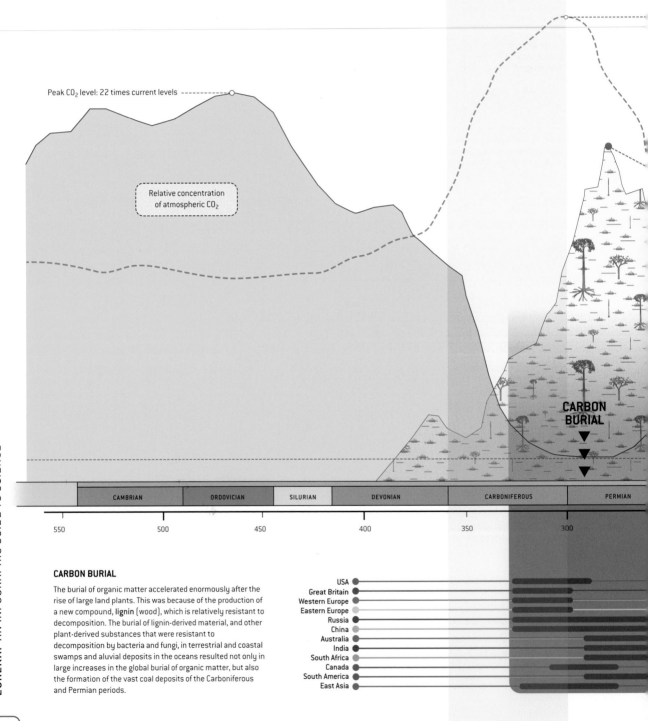

Peak CO_2 level: 22 times current levels

Relative concentration of atmospheric CO_2

CARBON BURIAL

| | CAMBRIAN | ORDOVICIAN | SILURIAN | DEVONIAN | CARBONIFEROUS | PERMIAN |

550   500   450   400   350   300

CARBON BURIAL

The burial of organic matter accelerated enormously after the rise of large land plants. This was because of the production of a new compound, **lignin** (wood), which is relatively resistant to decomposition. The burial of lignin-derived material, and other plant-derived substances that were resistant to decomposition by bacteria and fungi, in terrestrial and coastal swamps and aluvial deposits in the oceans resulted not only in large increases in the global burial of organic matter, but also the formation of the vast coal deposits of the Carboniferous and Permian periods.

USA
Great Britain
Western Europe
Eastern Europe
Russia
China
Australia
India
South Africa
Canada
South America
East Asia

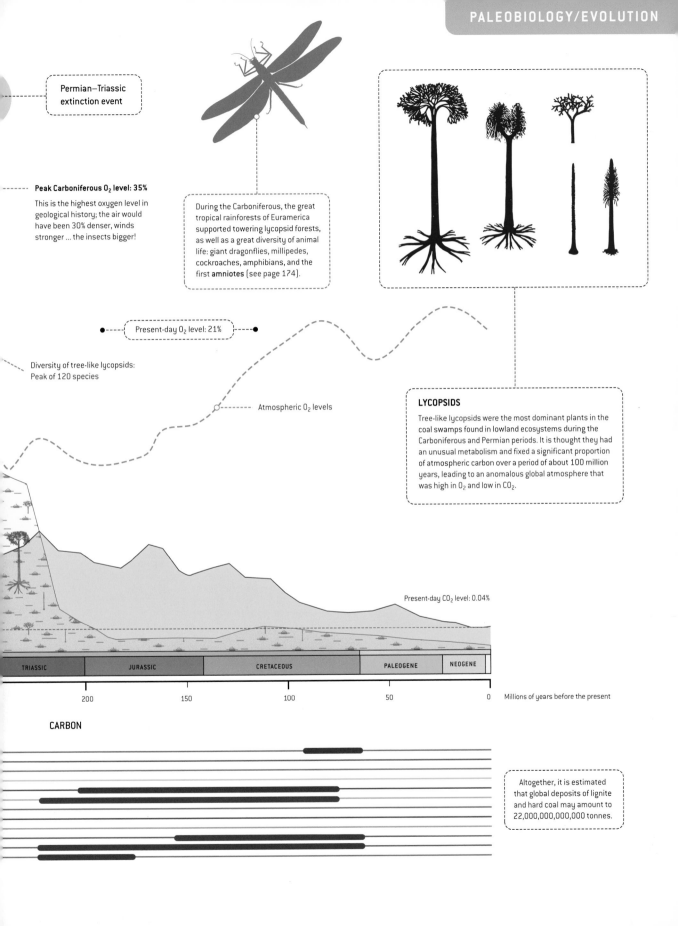

Permian–Triassic extinction event

Peak Carboniferous O_2 level: 35%

This is the highest oxygen level in geological history; the air would have been 30% denser, winds stronger ... the insects bigger!

During the Carboniferous, the great tropical rainforests of Euramerica supported towering lycopsid forests, as well as a great diversity of animal life: giant dragonflies, millipedes, cockroaches, amphibians, and the first **amniotes** (see page 174).

Present-day O_2 level: 21%

Diversity of tree-like lycopsids:
Peak of 120 species

Atmospheric O_2 levels

LYCOPSIDS

Tree-like lycopsids were the most dominant plants in the coal swamps found in lowland ecosystems during the Carboniferous and Permian periods. It is thought they had an unusual metabolism and fixed a significant proportion of atmospheric carbon over a period of about 100 million years, leading to an anomalous global atmosphere that was high in O_2 and low in CO_2.

Present-day CO_2 level: 0.04%

TRIASSIC	JURASSIC	CRETACEOUS	PALEOGENE	NEOGENE

200   150   100   50   0   Millions of years before the present

CARBON

Altogether, it is estimated that global deposits of lignite and hard coal may amount to 22,000,000,000,000 tonnes.

This wildly successful phylum accounts for over 80 percent of all known living animal species.

Meganeura monyi
- the largest known flying insect species – wingspan of up to 65 centimetres
- **Period:** Carboniferous
- **Temporal range:** 305–299 Ma

Jaekelopterus rhenaniae
- Extinct genus of sea scorpion – largest known arthropod
- **Period:** Middle Devonian
- **Temporal range:** 390 Ma (millions of years before the present)

Arthropleura armata
- Extinct millipede – largest known terrestrial arthropod
- **Period:** Late Carboniferous
- **Temporal range:** 323–299 Ma

AN ARTHROPOD
MENAGERIE

A selection of the largest know extinct arthropods, shown relative to the size of a human. During the Carboniferous period, very high atmospheric oxygen levels, and an absence of vertebrate predators, enabled land arthropods to grow to grotesque sizes by modern standards!

Isotelus rex
- Currently the largest known trilobite
- **Period:** Upper Ordovician
- **Temporal range:** 460–50 Ma

| CAMBRIAN | ORDOVICIAN | SILURIAN | DEVONIAN | CARBONIFEROUS |

Megaloprepus caerulatus
- largest living dragonfly
- wingspan: 19 centimetres

THE RELATIVE PROPORTIONS OF RECORED LIVING SPECIES

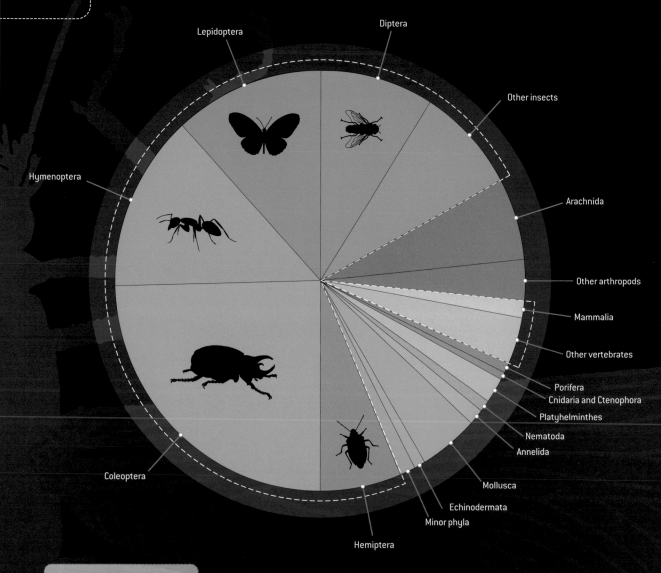

Lepidoptera

Diptera

Other insects

Hymenoptera

Arachnida

Other arthropods

Mammalia

Other vertebrates

Porifera
Cnidaria and Ctenophora
Platyhelminthes
Nematoda
Annelida

Mollusca

Echinodermata

Minor phyla

Coleoptera

Hemiptera

LIFE-SIZE MEGANEURA MONYI

In the background on this spread is a representation of the Carboniferous giant flying insect, life-size!

| ERMIAN | TRIASSIC | JURASSIC | CRETACEOUS | PALEOGENE | NEOGENE |

Amniotes

The amniotic egg represents a critical evolutionary divergence. It enabled life to reproduce on dry land, free of the need to return to water for reproduction. From this point the amniotes spread around the globe into every niche, eventually becoming the dominant land vertebrates.

OSTRICH

The largest eggs of any living bird

- **Length:** about 15 cm
- **Weight:** about 1.4 kg
- **Volume:** 24 x a chicken egg

AEPYORNIS

The largest type of bird egg ever found. *Aepyornis* is a genus of huge, extinct, flightless birds that were found in Madagascar. They were known as elephant birds.

- **Length:** up to 34 cm
- **Weight:** about 10 kg
- **Volume:** 160 x a chicken egg

BEE HUMMINGBIRD

The smallest known amniote egg.
- **Length:** up to 7 mm
- **Weight:** about 0.00025 kg
- **Volume:** 1 chicken egg equals 540 bee hummingbird eggs.

DINOSAUR EGG

SAUROPODS

Extinct clade of enormous herbivorous dinosaurs that included the well-known *Diplodocus*. They laid the largest known dinosaur eggs. Despite the animals' size, their eggs were not that much bigger than an ostrich egg.

- **Length:** about 18 cm
- **Weight:** over 2 kg
- **Volume:** 30 x a chicken egg

Mammals
Anapsids
Plesiosaurs
Ichthyosaurs
Sphenodontia
Lizards and snakes
Turtles
Crocodiles/alligators
Pterosaurs
Ornithischian dinosaurs
Saurischian dinosaurs
Birds

MAMMAL EGGS?

The earliest ancestral mammals laid eggs, as do the rather unusual platypus and echidna to this day. Other marsupial and placental mammals do not lay eggs, but their unborn young are surrounded by the same tissues that identify all amniotes. In placental mammals, the egg itself is void of yolk, but develops an umbilical cord from structures that in reptiles would form the yolk sac. Receiving nutrients from the mother, the foetus completes its development inside the uterus and is born **viviparously** (live-born).

??

THE AMNIOTE FAMILY TREE ▶

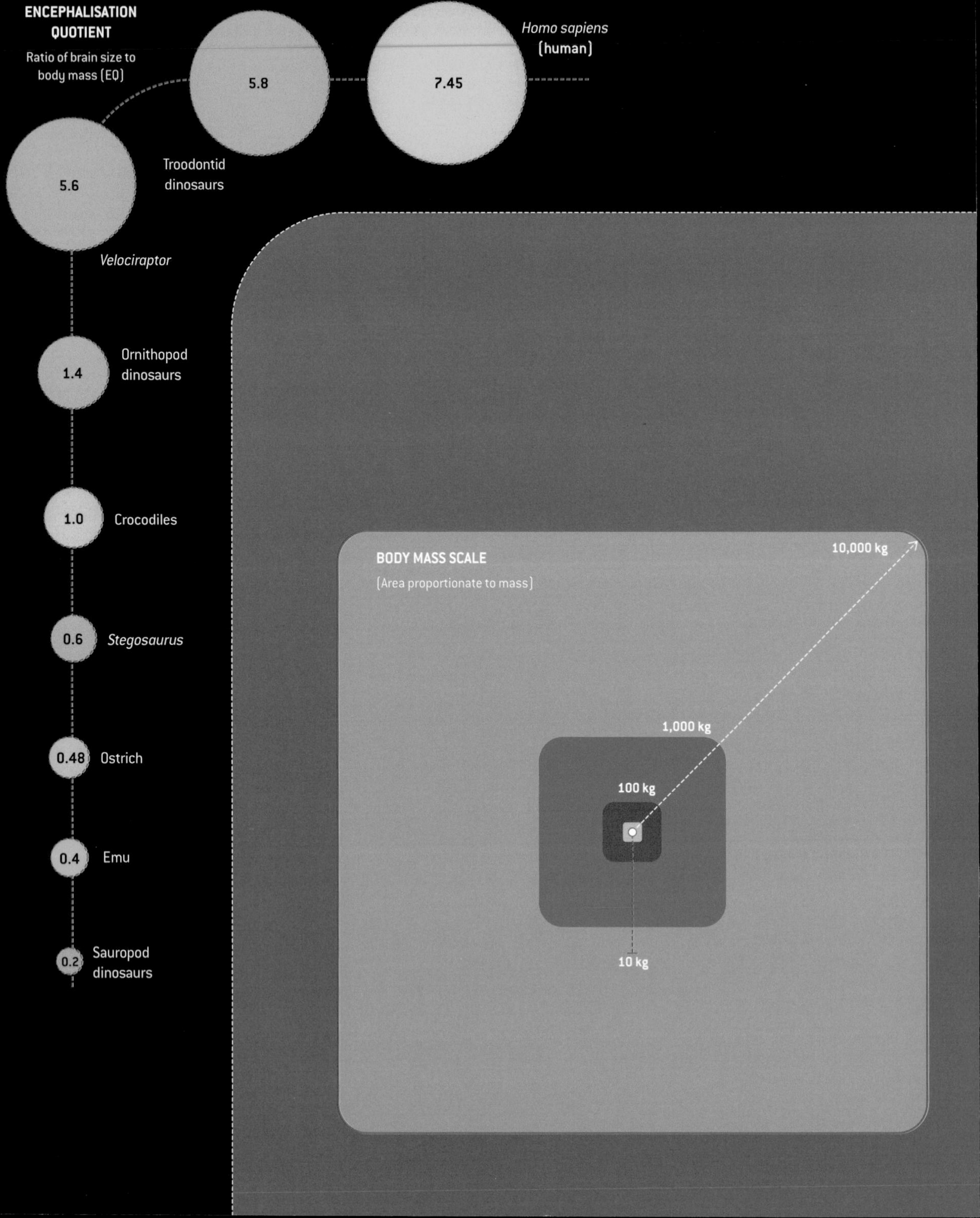

ENCEPHALISATION QUOTIENT
Ratio of brain size to body mass (EQ)
5.6
Velociraptor
5.8
Troodontid dinosaurs
7.45
Homo sapiens (human)
1.4
Ornithopod dinosaurs
1.0
Crocodiles
0.6
Stegosaurus
0.48
Ostrich
0.4
Emu
0.2
Sauropod dinosaurs
BODY MASS SCALE
(Area proportionate to mass)
10,000 kg
1,000 kg
100 kg
10 kg

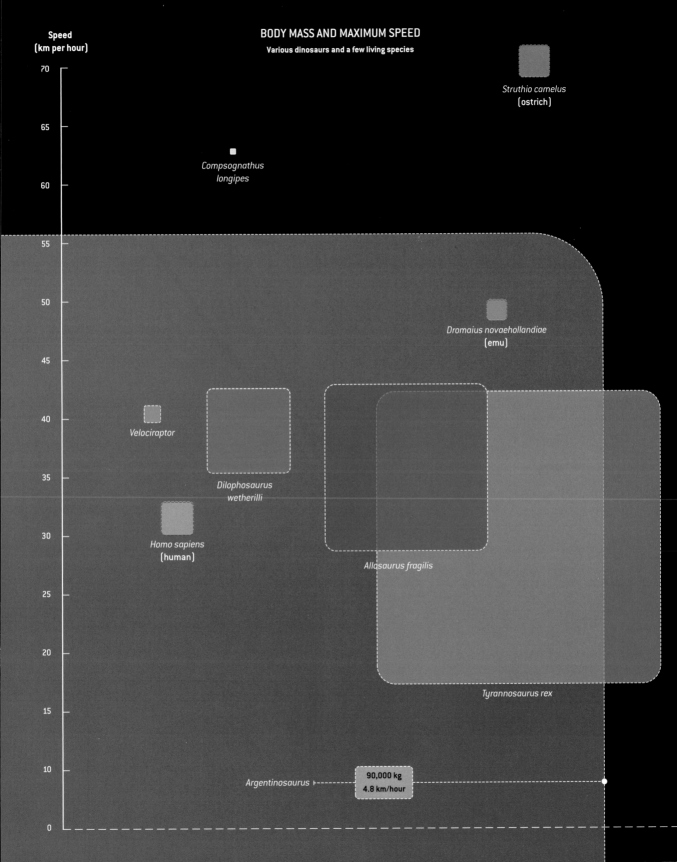

**Speed
(km per hour)**

BODY MASS AND MAXIMUM SPEED
Various dinosaurs and a few living species

Struthio camelus
(ostrich)

70

65

*Compsognathus
longipes*

60

55

50

Dromaius novaehollandiae
(emu)

45

Velociraptor

40

*Dilophosaurus
wetherilli*

35

Homo sapiens
(human)

30

Allosaurus fragilis

25

20

Tyrannosaurus rex

15

10

Argentinosaurus ├------- 90,000 kg
4.8 km/hour

5

0

From the emergence of the first true trees 350 million years ago, through the Carboniferous period when forests dominated the Earth's landmasses, to the present day when these lungs of our planet are under threat from humankind, trees have shaped our world.

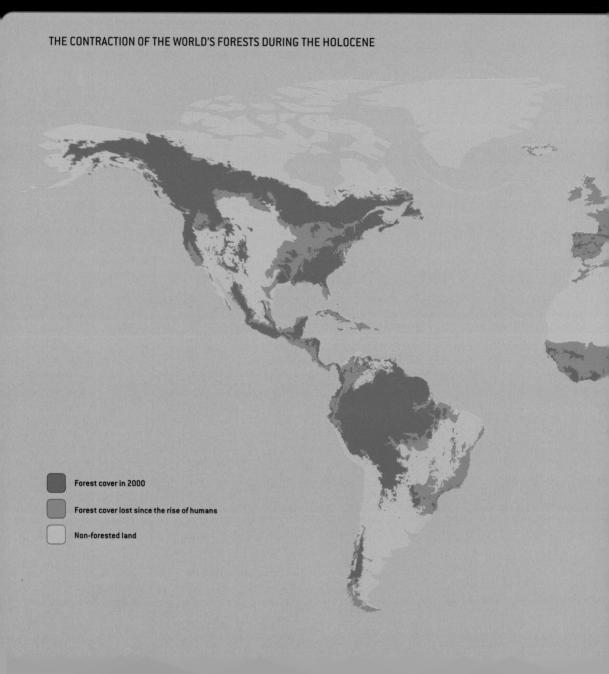

THE CONTRACTION OF THE WORLD'S FORESTS DURING THE HOLOCENE

Forest cover in 2000

Forest cover lost since the rise of humans

Non-forested land

THE BOREAL AGE

About 1,000 years after the end of the last Ice Age, some 10,000 years before present, global temperatures rose rapidly and forests spread, replacing the open grasslands and steppes. Forest replaced the open lands in Europe, and forest-dwelling animals, including humans, spread from their southerly refuges and replaced the ice-age mammals. Trees in the current epoch were at their peak — it is estimated that there were twice as many trees as there are today.

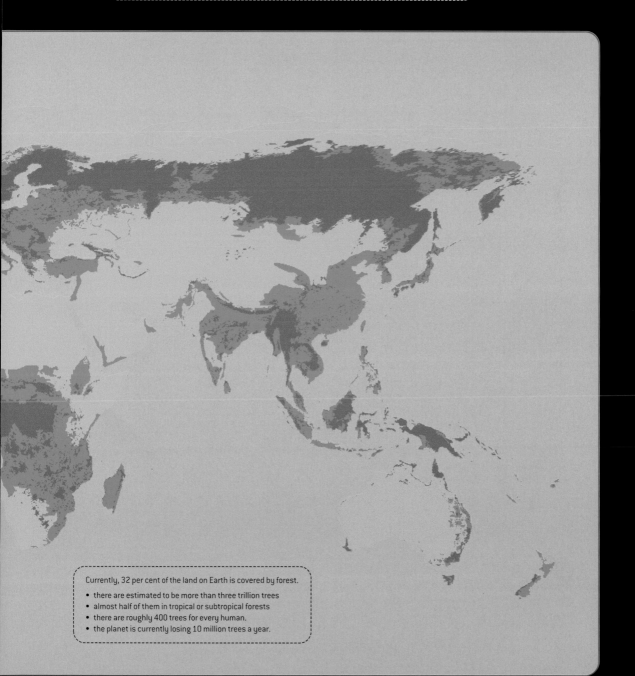

Currently, 32 per cent of the land on Earth is covered by forest.

- there are estimated to be more than three trillion trees
- almost half of them in tropical or subtropical forests
- there are roughly 400 trees for every human.
- the planet is currently losing 10 million trees a year.

Pterosaurs: vertebrate flight

The first vertebrates to evolve true flight were the pterosaurs, flying archosaurian reptiles. Modern science reveals them to be adept and powerful fliers, ruling the skies for over 160 million years ... longer than birds have been around.

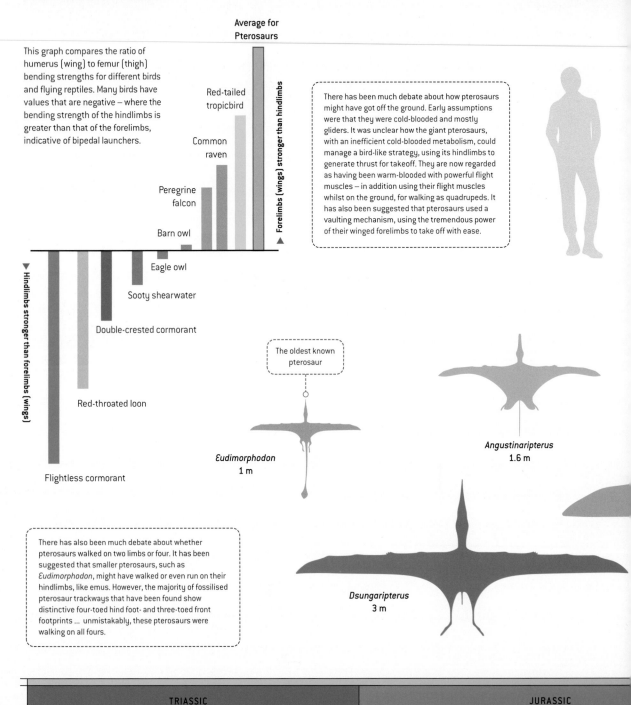

Average for Pterosaurs

This graph compares the ratio of humerus (wing) to femur (thigh) bending strengths for different birds and flying reptiles. Many birds have values that are negative – where the bending strength of the hindlimbs is greater than that of the forelimbs, indicative of bipedal launchers.

Red-tailed tropicbird

Common raven

Peregrine falcon

Barn owl

▲ Forelimbs (wings) stronger than hindlimbs

▼ Hindlimbs stronger than forelimbs (wings)

Eagle owl

Sooty shearwater

Double-crested cormorant

Red-throated loon

Flightless cormorant

There has been much debate about how pterosaurs might have got off the ground. Early assumptions were that they were cold-blooded and mostly gliders. It was unclear how the giant pterosaurs, with an inefficient cold-blooded metabolism, could manage a bird-like strategy, using its hindlimbs to generate thrust for takeoff. They are now regarded as having been warm-blooded with powerful flight muscles – in addition using their flight muscles whilst on the ground, for walking as quadrupeds. It has also been suggested that pterosaurs used a vaulting mechanism, using the tremendous power of their winged forelimbs to take off with ease.

The oldest known pterosaur

Eudimorphodon
1 m

Angustinaripterus
1.6 m

There has also been much debate about whether pterosaurs walked on two limbs or four. It has been suggested that smaller pterosaurs, such as *Eudimorphodon*, might have walked or even run on their hindlimbs, like emus. However, the majority of fossilised pterosaur trackways that have been found show distinctive four-toed hind foot- and three-toed front footprints ... unmistakably, these pterosaurs were walking on all fours.

Dsungaripterus
3 m

TRIASSIC	JURASSIC

250   Millions of years before the present   200

WINGSPAN AND ERAS OF NOTABLE PTEROSAURS

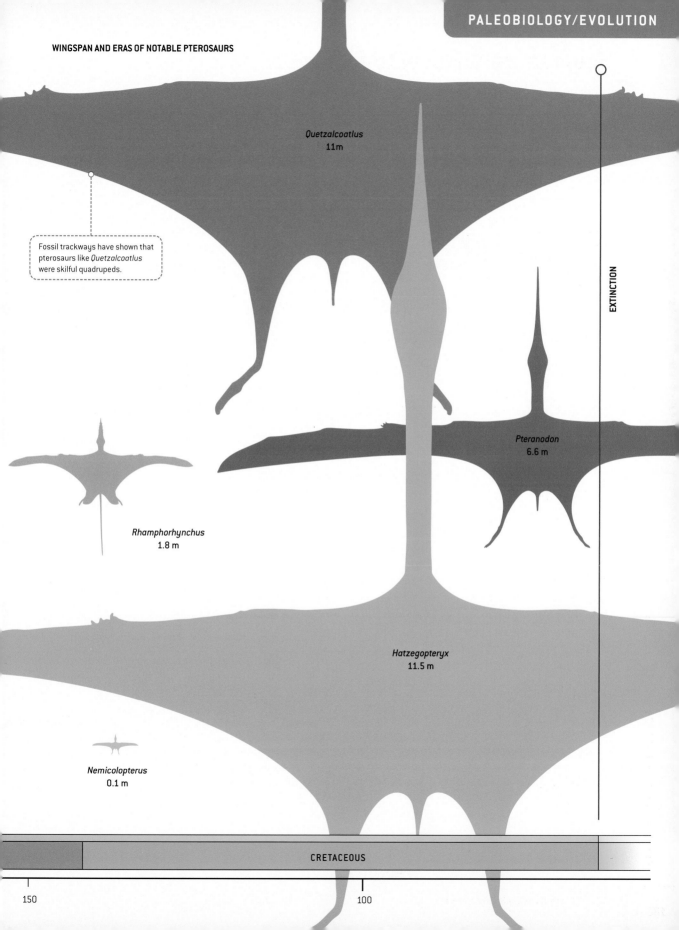

Quetzalcoatlus
11m

Fossil trackways have shown that
pterosaurs like *Quetzalcoatlus*
were skilful quadrupeds.

EXTINCTION

Pteranodon
6.6 m

Rhamphorhynchus
1.8 m

Hatzegopteryx
11.5 m

Nemicolopterus
0.1 m

CRETACEOUS

150

100

Angiosperms and flying insects

Angiosperms, flowering plants, exploded in diversity at the end of the Cretaceous period from about 180 genera to the more than 14,000 modern genera. This incredible success is founded on coevolution with flying insects.

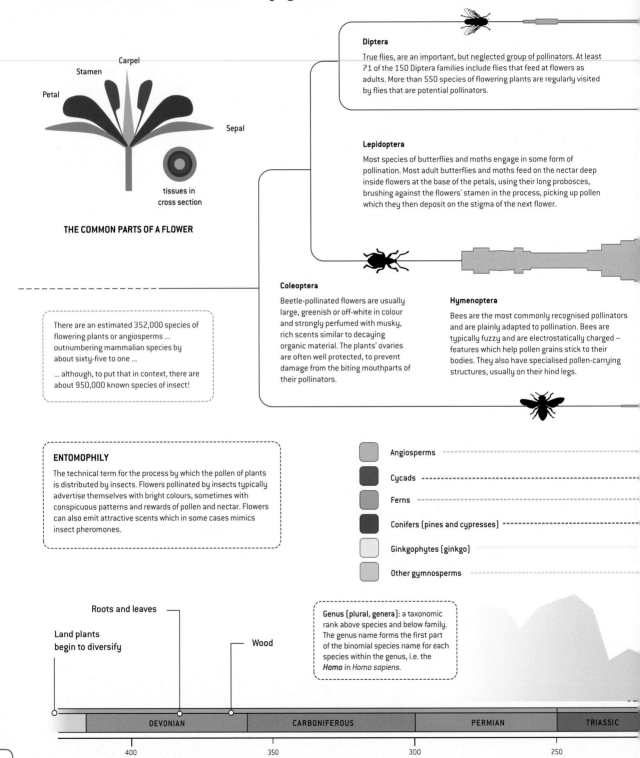

Diptera

True flies, are an important, but neglected group of pollinators. At least 71 of the 150 Diptera families include flies that feed at flowers as adults. More than 550 species of flowering plants are regularly visited by flies that are potential pollinators.

Lepidoptera

Most species of butterflies and moths engage in some form of pollination. Most adult butterflies and moths feed on the nectar deep inside flowers at the base of the petals, using their long probosces, brushing against the flowers' stamen in the process, picking up pollen which they then deposit on the stigma of the next flower.

Coleoptera

Beetle-pollinated flowers are usually large, greenish or off-white in colour and strongly perfumed with musky, rich scents similar to decaying organic material. The plants' ovaries are often well protected, to prevent damage from the biting mouthparts of their pollinators.

Hymenoptera

Bees are the most commonly recognised pollinators and are plainly adapted to pollination. Bees are typically fuzzy and are electrostatically charged – features which help pollen grains stick to their bodies. They also have specialised pollen-carrying structures, usually on their hind legs.

Carpel
Stamen
Petal
Sepal

tissues in cross section

THE COMMON PARTS OF A FLOWER

There are an estimated 352,000 species of flowering plants or angiosperms ... outnumbering mammalian species by about sixty-five to one ...

... although, to put that in context, there are about 950,000 known species of insect!

ENTOMOPHILY

The technical term for the process by which the pollen of plants is distributed by insects. Flowers pollinated by insects typically advertise themselves with bright colours, sometimes with conspicuous patterns and rewards of pollen and nectar. Flowers can also emit attractive scents which in some cases mimics insect pheromones.

Angiosperms

Cycads

Ferns

Conifers (pines and cypresses)

Ginkgophytes (ginkgo)

Other gymnosperms

Roots and leaves

Land plants begin to diversify

Wood

Genus (plural, genera): a taxonomic rank above species and below family. The genus name forms the first part of the binomial species name for each species within the genus, i.e. the *Homo* in *Homo sapiens*.

| DEVONIAN | CARBONIFEROUS | PERMIAN | TRIASSIC |

400   350   300   250

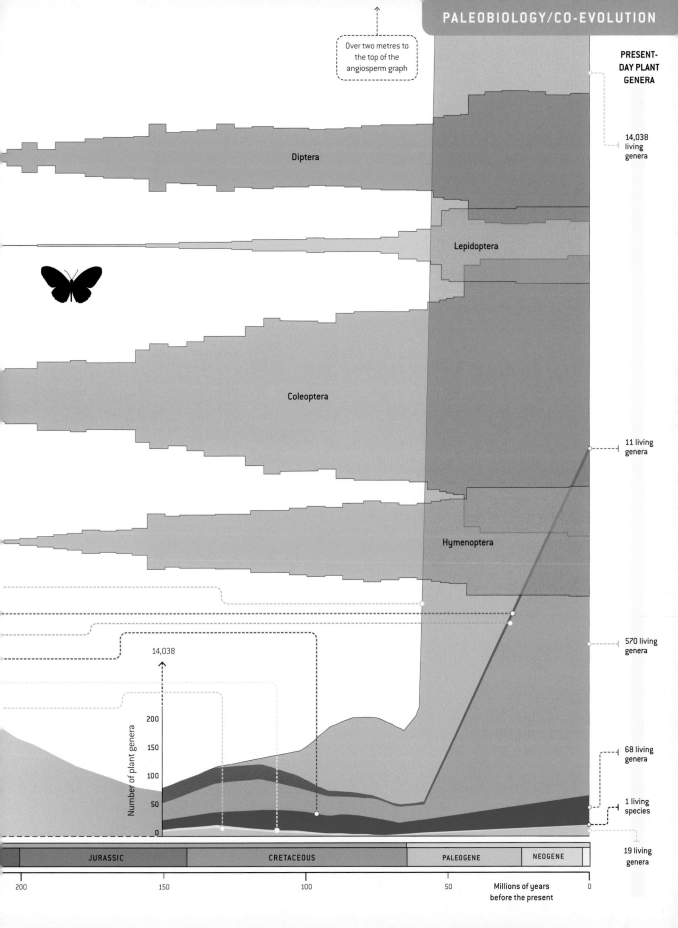

Over two metres to the top of the angiosperm graph

PRESENT-DAY PLANT GENERA

14,038 living genera

Diptera

Lepidoptera

Coleoptera

Hymenoptera

11 living genera

570 living genera

68 living genera

1 living species

19 living genera

14,038

Number of plant genera

200

150

100

50

JURASSIC

CRETACEOUS

PALEOGENE

NEOGENE

200

150

100

50

Millions of years before the present

0

Vertebrate vision

For over 500 million years, vision has been a perceptual system shared by almost all animals. Invertebrate vision peaks with the insect compound eye, with its rich colour sense, and the extraordinarily refined vision of the octopus and squid. But in vertebrates, seeing reaches new levels of richness and detail.

1°
Human vision can resolve 40 to 60 bars within each degree of visual field

An index finger subtends about 1° at arm's length

ACUITY

Crab-eating macaque
Chimpanzee
Human
Skylark
Magpie
American kestrel
Freshwater turtle
Pigeon
Barn owl
Starling
Woodpecker
House fly
Cat
Bactrian camel
Bottlenose dolphin (air)
Bottlenose dolphin (water)
Horse
Killer whale
European robin
Lynx
Cattle
Goldfish
Dog
Jackdaw
Magpie
Skylark
Capuchin monkey
Australian wedge-tail eagle

100 90 80 70 60 50 40 30 20 10

Bars resolvable per degree

Acuity, the abilty to resolve fine detail, varies enormously from species to species. For some animals, such as antelope, low acuity with a wide field of view and great sensitivity to movement, is what keeps you alive. For airborne raptors, the ability to sense infinitesimal movement at distance is the thing. For humans, with our rich world of tools, mind and intelligent manipulation, good acuity allied with excellent stereoscopic colour vision enables us to visually sense a world that matches our ability to conceive it.

STEREOSCOPIC VISION

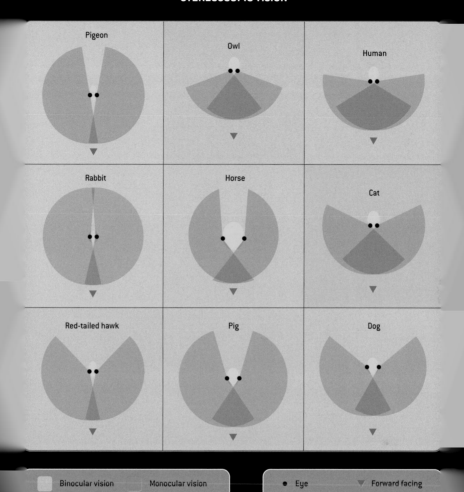

Pigeon

Owl

Human

Rabbit

Horse

Cat

Red-tailed hawk

Pig

Dog

Binocular vision   Monocular vision   ● Eye   ▼ Forward facing

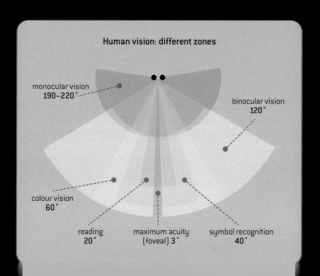

Human vision: different zones

monocular vision
190–220°

binocular vision
120°

colour vision
60°

reading
20°

maximum acuity
(foveal) 3°

symbol recognition
40°

Insect vision?

To achieve a resolution comparable to our foveal acuity, humans would require very large compound eyes of the sort found in insects. They'd have to be about 22 metres in diameter.

The Chicxulub Impactor

A huge meteorite or comet hit the Earth off the coast of present-day Mexico, around 65.5 million years ago. It is now thought this event triggered one of the greatest extinctions of life on Earth, bringing to an abrupt end the reign of the dinosaurs.

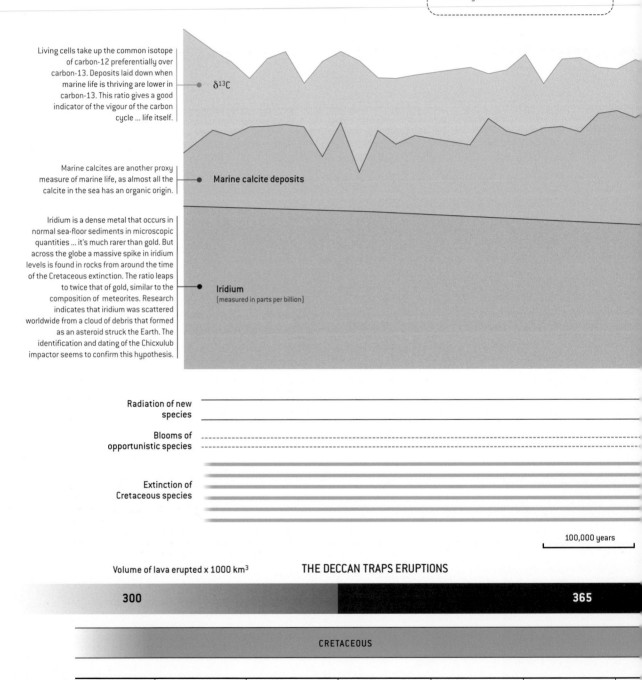

Living cells take up the common isotope of carbon-12 preferentially over carbon-13. Deposits laid down when marine life is thriving are lower in carbon-13. This ratio gives a good indicator of the vigour of the carbon cycle ... life itself.

$\delta^{13}C$

Marine calcites are another proxy measure of marine life, as almost all the calcite in the sea has an organic origin.

Marine calcite deposits

Iridium is a dense metal that occurs in normal sea-floor sediments in microscopic quantities ... it's much rarer than gold. But across the globe a massive spike in iridium levels is found in rocks from around the time of the Cretaceous extinction. The ratio leaps to twice that of gold, similar to the composition of meteorites. Research indicates that iridium was scattered worldwide from a cloud of debris that formed as an asteroid struck the Earth. The identification and dating of the Chicxulub impactor seems to confirm this hypothesis.

Iridium
(measured in parts per billion)

Radiation of new species

Blooms of opportunistic species

Extinction of Cretaceous species

100,000 years

Volume of lava erupted x 1000 km³

THE DECCAN TRAPS ERUPTIONS

300

365

CRETACEOUS

66.2   66.1   66   65.9   65.8   65.7

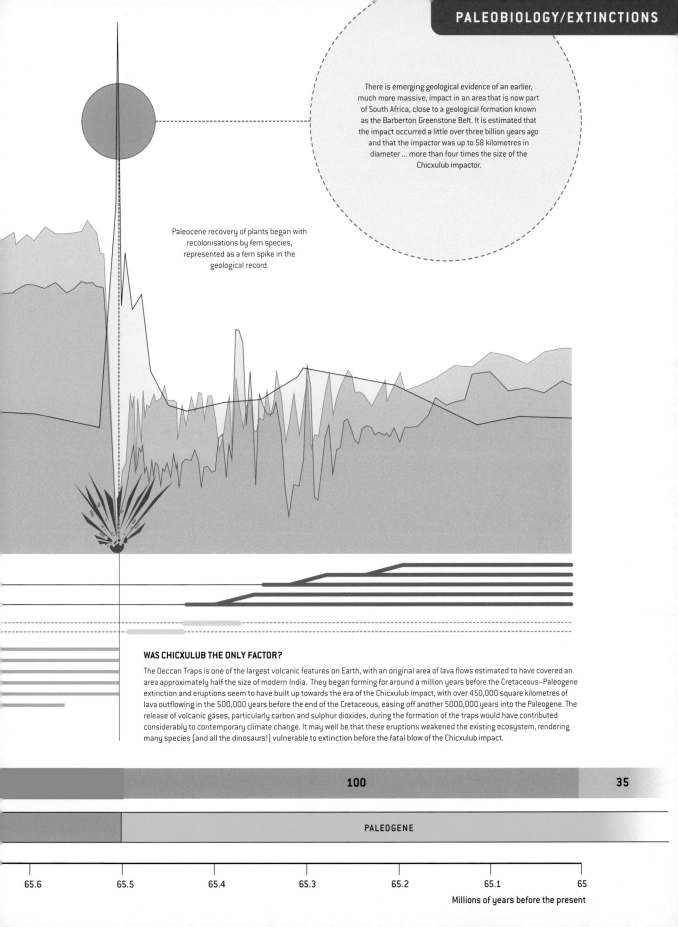

There is emerging geological evidence of an earlier, much more massive, impact in an area that is now part of South Africa, close to a geological formation known as the Barberton Greenstone Belt. It is estimated that the impact occurred a little over three billion years ago and that the impactor was up to 58 kilometres in diameter ... more than four times the size of the Chicxulub impactor.

Paleocene recovery of plants began with recolonisations by fern species, represented as a fern spike in the geological record.

WAS CHICXULUB THE ONLY FACTOR?

The Deccan Traps is one of the largest volcanic features on Earth, with an original area of lava flows estimated to have covered an area approximately half the size of modern India. They began forming for around a million years before the Cretaceous–Paleogene extinction and eruptions seem to have built up towards the era of the Chicxulub impact, with over 450,000 square kilometres of lava outflowing in the 500,000 years before the end of the Cretaceous, easing off another 5000,000 years into the Paleogene. The release of volcanic gases, particularly carbon and sulphur dioxides, during the formation of the traps would have contributed considerably to contemporary climate change. It may well be that these eruptions weakened the existing ecosystem, rendering many species (and all the dinosaurs!) vulnerable to extinction before the fatal blow of the Chicxulub impact.

| 100 | 35 |

PALEOGENE

| 65.6 | 65.5 | 65.4 | 65.3 | 65.2 | 65.1 | 65 |

Millions of years before the present

Bird migration

Birds were the sole surviving group of dinosaur descendants after the Cretaceous mass extinction. Today there are nearly 10,000 living species and the power of flight enables them to inhabit niches that involve often Odyssean biannual migrations from overwintering to breeding sites and the return journeys all over the world on well-established routes or flyways.

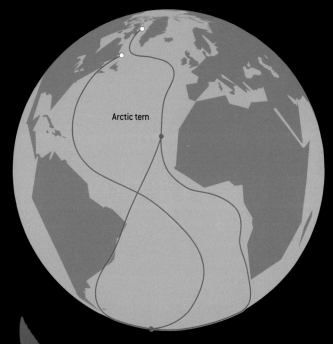

Arctic tern

Osprey
Pandion haliaetus
A large bird of prey with a wingspan that can reach 180 cm. It is found in temperate and tropical regions throughout all the continents, excepting Antarctica. European birds spend the summer throughout the north of the continent, into Ireland, Scandinavia and the British Isles, though not Iceland, and winters in North Africa.

Arctic tern _Sterna paradisaea_
Strongly migratory, living through two summers each year as it migrates along a convoluted route from its northern breeding grounds to the Antarctic coast for the southern summer and back again six months later. Recent studies have shown average annual round-trip lengths of between 80,000 and 90,000 km, by far the longest migrations known in the animal kingdom.

Bobolink _Dolichonyx oryzivorus_
Breeds in the summer in North America across much of southern Canada and the northern United States. It migrates long distances, overwintering in southern South America in Argentina, Bolivia, Brazil and Paraguay. Individuals have been tracked migrating 19,000 km over the course of the year, flying up to 1,800 km in a single day.

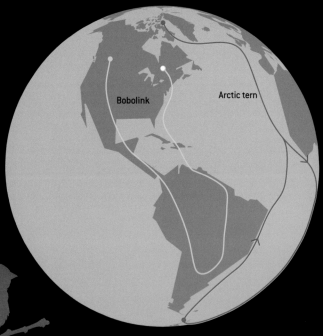

Bobolink

Arctic tern

Bar-tailed godwit *Limosa lapponica*

A large wader which breeds on Arctic coasts and tundra and winters on coasts in temperate and tropical regions, including Australia and New Zealand. Has been recorded making the longest known non-stop flight of any migrant, flying 11,700 km from Alaska to New Zealand in eight days. Prior to migration, 55 percent of their bodyweight is stored as fat in order to fuel this uninterrupted journey.

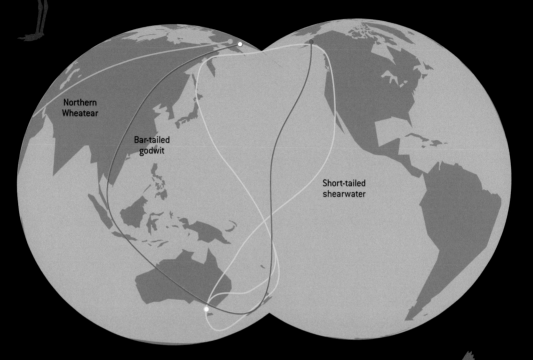

Northern Wheatear

Bar-tailed godwit

Short-tailed shearwater

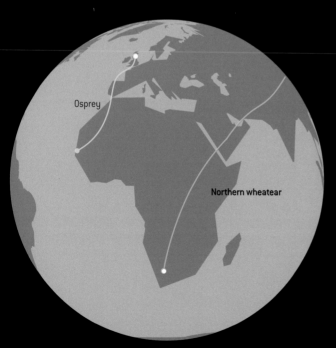

Osprey

Northern wheatear

Short-tailed shearwater *Ardenna tenuirostris*

Another epic migratory seabird, which breeds mainly on small islands in Bass Strait and Tasmania and migrates to the seas off the Aleutian Islands and Kamchatka. In the northern autumn they travel down the coast of California before crossing the Pacific back to Australia. Historically known as the Muttonbird, due to its prized flesh.

Northern wheatear *Oenanthe oenanthe*

The northern wheatear is a migratory insectivorous species that breeds in northern Europe and Asia with footholds in northern Canada and Greenland. Their migratory routes lead to overwintering territories in Sub-Saharan Africa – a journey that can exceed 30,000 km.

The rise of the mammals

Following the Cretaceous–Paleogene mass extinction and the demise of the dinosaurs, mammals blossomed, diversified and grew in size to dominate the majority of the planet's ecosystems.

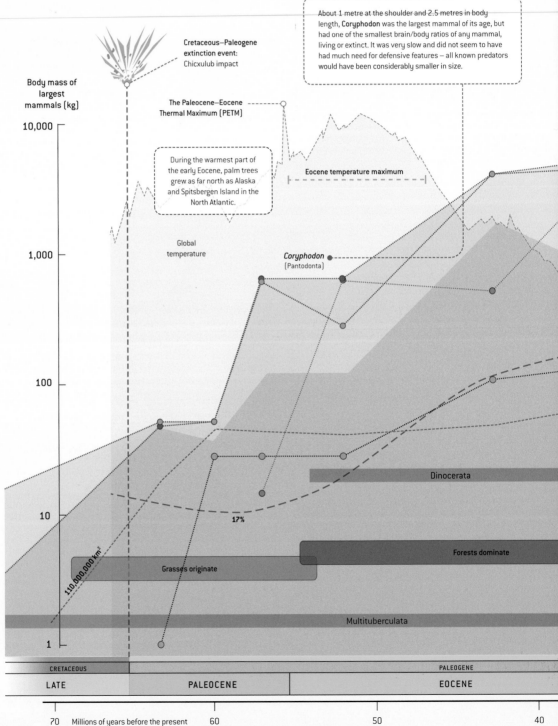

About 1 metre at the shoulder and 2.5 metres in body length, **Coryphodon** was the largest mammal of its age, but had one of the smallest brain/body ratios of any mammal, living or extinct. It was very slow and did not seem to have had much need for defensive features – all known predators would have been considerably smaller in size.

Cretaceous–Paleogene extinction event: Chicxulub impact

Body mass of largest mammals (kg)

The Paleocene–Eocene Thermal Maximum (PETM)

10,000

During the warmest part of the early Eocene, palm trees grew as far north as Alaska and Spitsbergen Island in the North Atlantic.

Eocene temperature maximum

Global temperature

1,000

Coryphodon
(Pantodonta)

100

Dinocerata

17%

110,000,000 km²

Forests dominate

10

Grasses originate

Multituberculata

1

CRETACEOUS | PALEOGENE

LATE | PALEOCENE | EOCENE

70   Millions of years before the present   60   50   40

Paraceratherium is an extinct genus of hornless rhinoceros, and amongst the largest terrestrial mammals ever to have existed at about 4.8 metres at the shoulder and 7.4 metres in length. It became extinct after surviving for about 11 million years for reasons unknown, but probably multiple: possibly climate change and the influx of elephant-like proboscideans.

Body mass of largest mammalian species on each continent: green tint represents **herbivores**, the lower brown tint the **heaviest carnivores**.

Proboscidea
Mammoths, American mastodons and Asian straight-tusked elephant, became the dominant land mammals during the late Miocene and into the Pleistocene.

Paraceratherium
(Perissodactyla)

Eurasia

Africa

North America

South America

Antarctic thawing

152,000,000 km²

Antarctic reglaciation

Continental land area

Atmospheric oxygen levels

21%

Antarctic Glaciation

Dinocerata: extinct plant-eating, rhinoceros-like hoofed mammals, known for their paired horns and tusk-like canine teeth.

C4 grasses expand

C4 grasses originate

C3 grasses increase

C4 grasses are able to thrive in warm, dry habitats with high amounts of sunlight. Their evolution led to the massive spread of the savannah. See page 134 for an explanation of C3 and C4 plants.

Multituberculata: extinct mouse-sized to beaver-sized rodent-like mammals which existed for about 120 million years and are often considered the most successful, diversified and long-lasting mammals.

	NEOGENE		
OLIGOCENE	MIOCENE	PLIOCENE	

30   20   10   0

Gestation

The length of pregnancy generally follows a close correlation to the size of the animal and brain size.

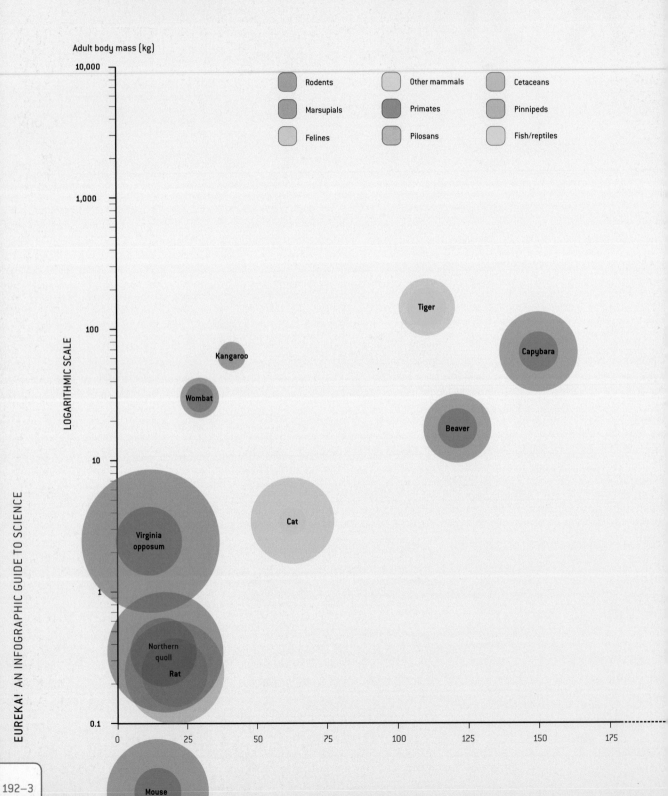

Adult body mass (kg)

LOGARITHMIC SCALE

10,000

1,000

100

10

1

0.1

Rodents
Marsupials
Felines
Other mammals
Primates
Pilosans
Cetaceans
Pinnipeds
Fish/reptiles

Tiger

Capybara

Kangaroo

Wombat

Beaver

Cat

Virginia opposum

Northern quoll

Rat

Mouse

0   25   50   75   100   125   150   175

LITTER SIZE

1

2

9

25

Sperm whale

Killer whale

African elephant

Hippopotamus

Walrus

Javan rhinoceros

Dolphin

Donkey

Seal

Sea lion

Gorilla

Human

Chimpanzee

Frilled shark

1277 days

Hoffmann's sloth

NON-MAMMALIAN VIVIPARY

The ability to give birth to live offspring evolved independently in certain animals, such as some arthropods, sharks and reptiles. Live birth in non-mammalian species is very rare. In most cases an identifiable egg is retained inside the mother up to the time of birth, but certain skinks, such as *Tiliqua* and *Corucia*, actually have a placenta that is attached directly to the mother — this is called viviparous matrotrophy.

Solomon Islands skink

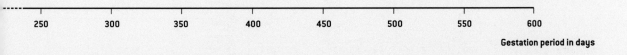

250   300   350   400   450   500   550   600

Gestation period in days

Evolutionary adaptation and return

Evolutionary radiation and diversification has led to most classes of animals conquering land and sea ... and some the skies above. The same process has led some to return to the bosom of the ocean.

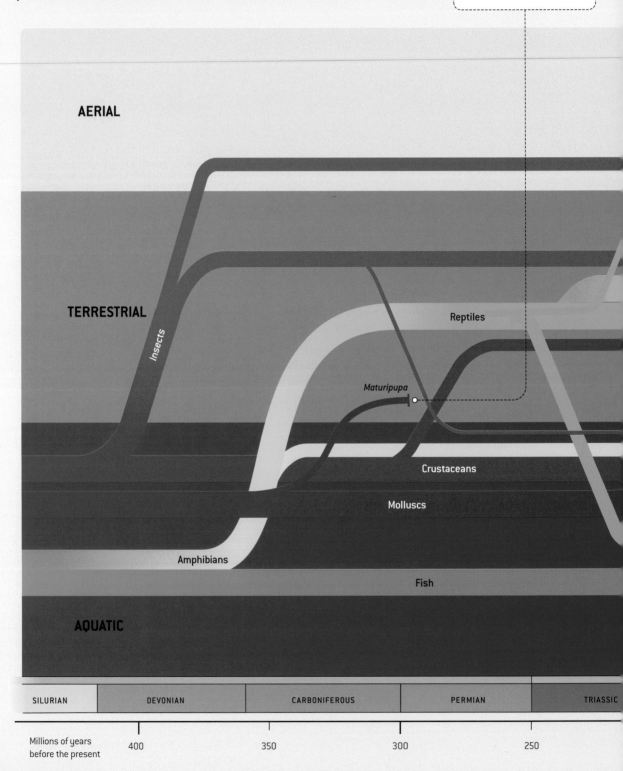

Maturipupa: earliest known terrestrial gastropods, found in the coal measures of the Carboniferous, but not directly related to the modern pulmonate land snails.

AERIAL

TERRESTRIAL

Insects

Reptiles

Maturipupa

Crustaceans

Molluscs

Amphibians

Fish

AQUATIC

| SILURIAN | DEVONIAN | CARBONIFEROUS | PERMIAN | TRIASSIC |

Millions of years before the present

400   350   300   250

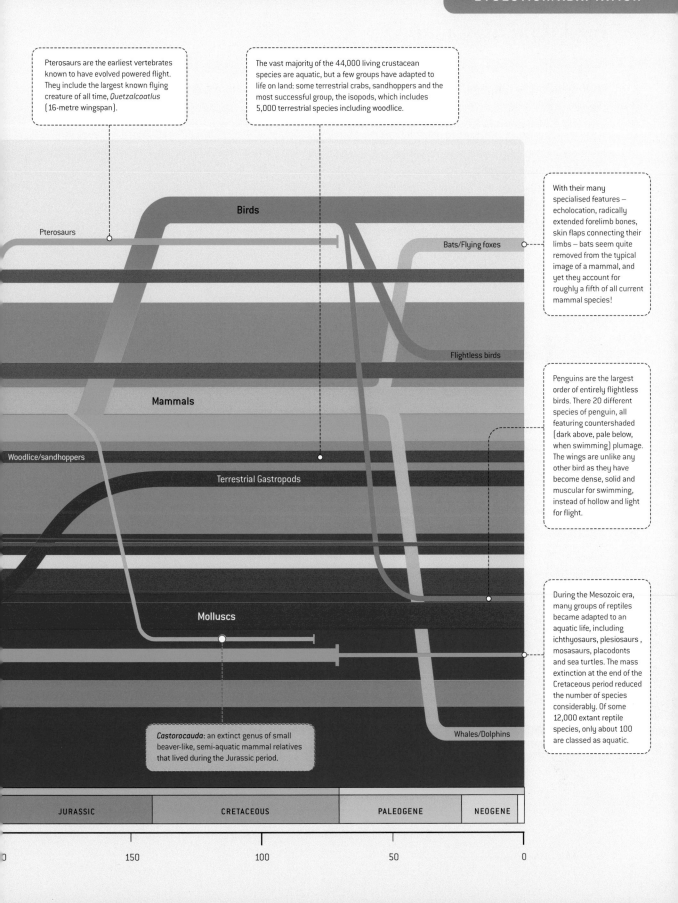

Pterosaurs are the earliest vertebrates known to have evolved powered flight. They include the largest known flying creature of all time, *Quetzalcoatlus* (16-metre wingspan).

The vast majority of the 44,000 living crustacean species are aquatic, but a few groups have adapted to life on land: some terrestrial crabs, sandhoppers and the most successful group, the isopods, which includes 5,000 terrestrial species including woodlice.

With their many specialised features — echolocation, radically extended forelimb bones, skin flaps connecting their limbs — bats seem quite removed from the typical image of a mammal, and yet they account for roughly a fifth of all current mammal species!

Penguins are the largest order of entirely flightless birds. There 20 different species of penguin, all featuring countershaded (dark above, pale below, when swimming) plumage. The wings are unlike any other bird as they have become dense, solid and muscular for swimming, instead of hollow and light for flight.

During the Mesozoic era, many groups of reptiles became adapted to an aquatic life, including ichthyosaurs, plesiosaurs, mosasaurs, placodonts and sea turtles. The mass extinction at the end of the Cretaceous period reduced the number of species considerably. Of some 12,000 extant reptile species, only about 100 are classed as aquatic.

Castorocauda: an extinct genus of small beaver-like, semi-aquatic mammal relatives that lived during the Jurassic period.

Birds

Pterosaurs

Bats/Flying foxes

Flightless birds

Mammals

Woodlice/sandhoppers

Terrestrial Gastropods

Molluscs

Whales/Dolphins

| JURASSIC | CRETACEOUS | PALEOGENE | NEOGENE |

150   100   50   0

Evolution of grasses

The evolution and subsequent ecological expansion of grasses since the Late Cretaceous have resulted in the establishment of one of Earth's dominant biomes, the temperate and tropical grasslands. As these verdant, open spaces spread, they provided the environment, and the new selective pressures, that eventually led to the rise of the primates.

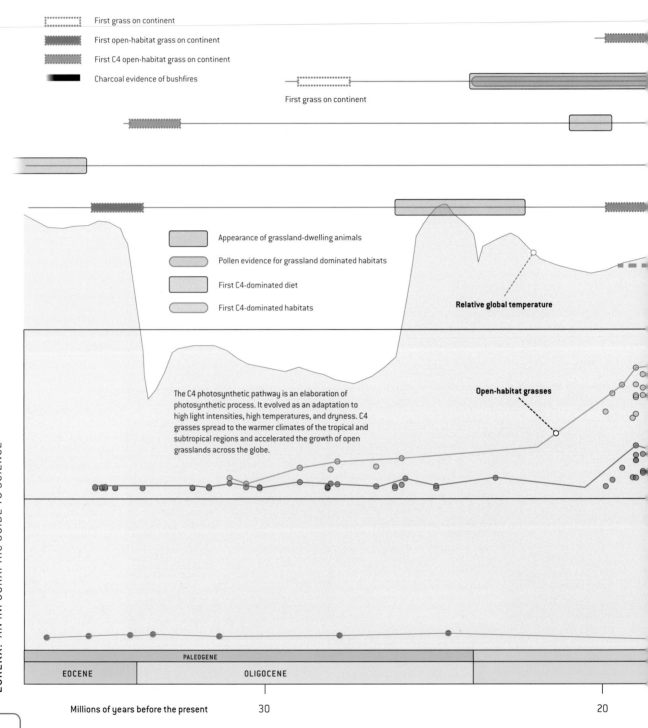

First grass on continent

First open-habitat grass on continent

First C4 open-habitat grass on continent

Charcoal evidence of bushfires

First grass on continent

Appearance of grassland-dwelling animals

Pollen evidence for grassland dominated habitats

First C4-dominated diet

First C4-dominated habitats

Relative global temperature

The C4 photosynthetic pathway is an elaboration of photosynthetic process. It evolved as an adaptation to high light intensities, high temperatures, and dryness. C4 grasses spread to the warmer climates of the tropical and subtropical regions and accelerated the growth of open grasslands across the globe.

Open-habitat grasses

PALEOGENE

EOCENE

OLIGOCENE

Millions of years before the present   30   20

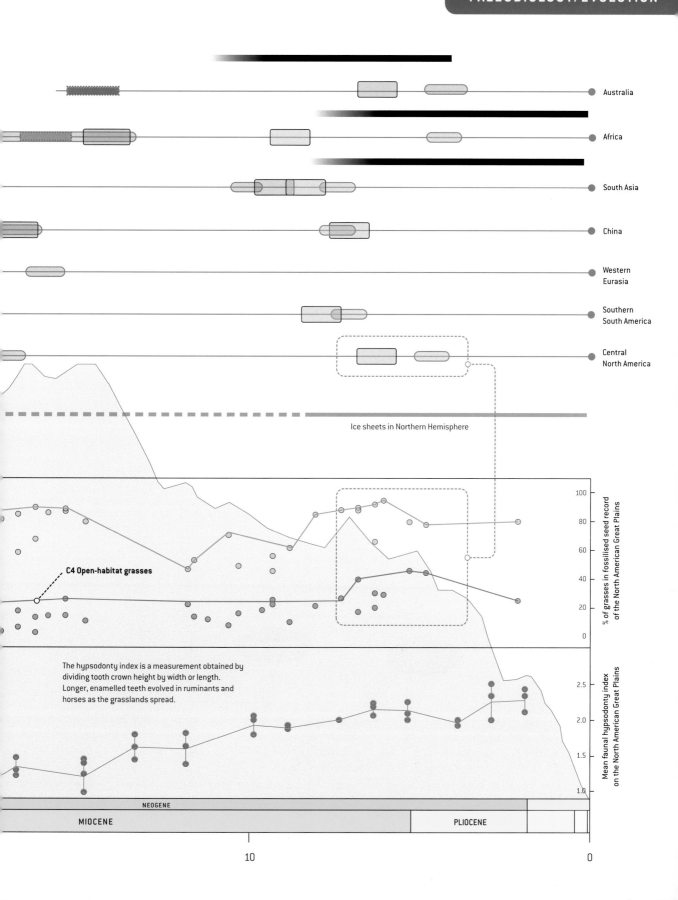

Australia

Africa

South Asia

China

Western
Eurasia

Southern
South America

Central
North America

Ice sheets in Northern Hemisphere

% of grasses in fossilised seed record
of the North American Great Plains

C4 Open-habitat grasses

100

80

60

40

20

0

The hypsodonty index is a measurement obtained by
dividing tooth crown height by width or length.
Longer, enamelled teeth evolved in ruminants and
horses as the grasslands spread.

Mean faunal hypsodonty index
on the North American Great Plains

2.5

2.0

1.5

1.0

NEOGENE

MIOCENE

PLIOCENE

10

0

Biosphere and the carbon cycle

The biosphere is the global ecological system integrating all living beings and their relationships, including their interaction with the elements of the lithosphere, geosphere, hydrosphere, pedosphere and atmosphere.

The pedosphere

The outermost layer of the Earth – the soil, or rather, the sum total of all the organisms, soils, water and air contained in the skin of the Earth. It is the foundation of all terrestrial life.

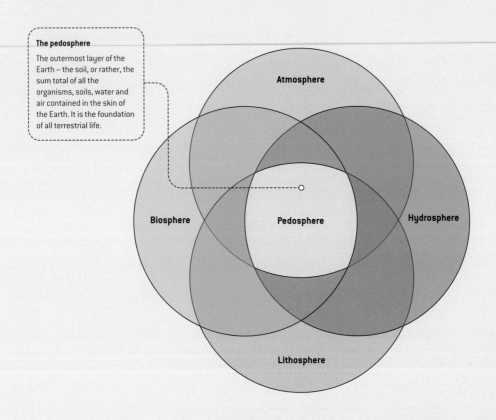

THE CARBON CYCLE

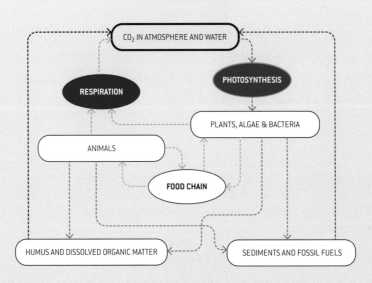

LIFE, THE CARBON CYCLE AND CLIMATE CHANGE

Carbon atoms are incorporated into living organic molecules by the photosynthetic activity of bacteria, algae and plants. They pass to microorganisms and animals in cyclical paths of eating and consumption; and into the soil, fresh water and oceans by excretion and death. Carbon is released back into the atmosphere as CO_2 when those organic molecules are oxidised by bacterial decompostion ... or when burned by humans.

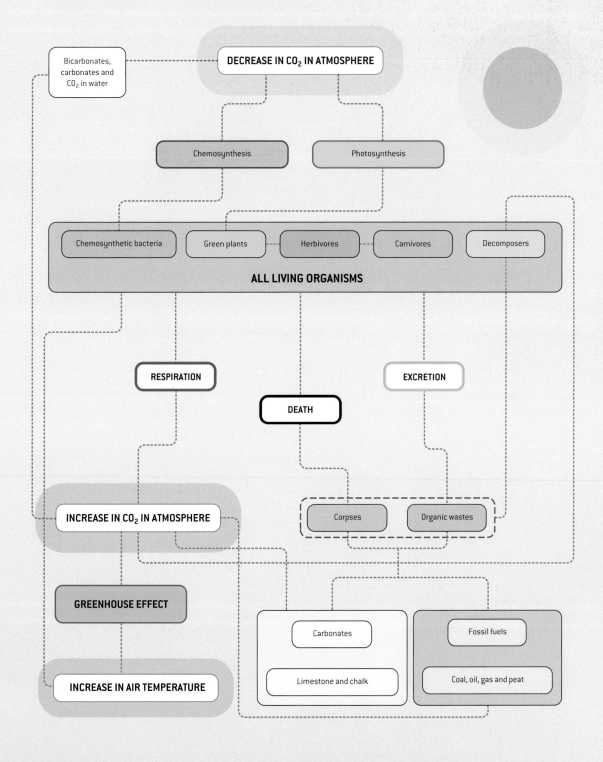

WoollyRhinoAAAGCAAGGCATTC
GGCTGAATTGiantLemurTTCCT
GGGTATATTACGGTTCATTHom
AAAACCCCCCCTCCCCCGCTTCTG
CAAACCCCAAAmericanMasto
GAGAAGACCCTGCGGAGCTTCAATT
GCTGAATTATCCGGTATCTACACGC
CAACTATGGCTGNATCANTCGATACA
TCTGCCTATTCATGiantShort-Face
GAGAGAAGGTTTGTGATGACTGTGI
TATATACACGCAAACGGAGCATCAATA
GACGAGGCCTATACTACGGATCGTATA
TAATTSabre-ToothedCatAGAACA

AAATGCCTAGATGAGGCGGTATATA

GCCTATTCATTCATGTTGGCCGAG

Sapiens Neanderthalensis

CCACAGCACTTAAACACATCTCTGC

on AACGGACATGAATTAATAAGAC

ACTAGTACAACT **Cave Lion** CTTATG

ACGGA **Blue Antelope** AGAGATGT

ACATGCNAACGCGGCATCAATATTCT

Bear GGCTACGTCCTACCATGAGGCT

sh **Elk** AATTACGGCTGAATTATCCGA

TCTTCATCTGCCTATTCATGCATGTAG

TTTTCTAGAAACATGAAACATTGGAG

CTCCTCTAGAGGGGATGTAAAGCACCG

Mitochondrial RNA base-pair sequences from extinct mammalian species

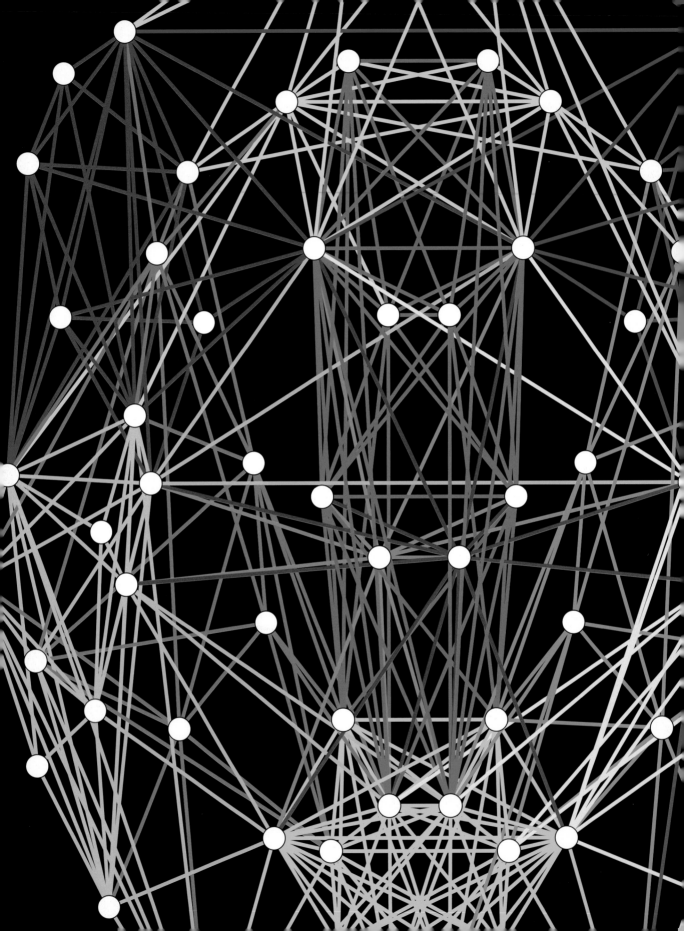

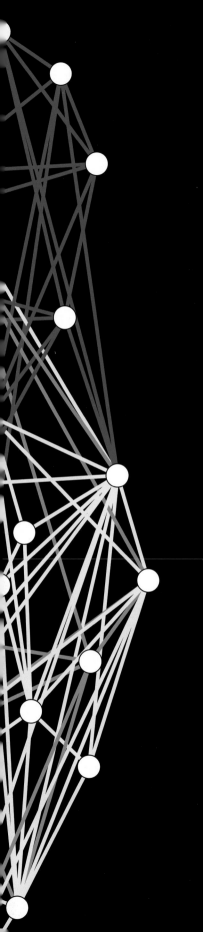

HUMAN

Human

The extinction event that marked the end of the Cretaceous period, 66 million years ago, famously drove the final nail in the coffin of the dinosaurs and is estimated to have led to the death of more than three-quarters of all the plant and animal species on Earth. This destruction, which had happened on a similar scale several times before in the long history of life, provided great evolutionary opportunities in its wake. With the disappearance of top predators, and the emptying of the landscape, many groups underwent rapid diversification – a sudden and prolific divergence into new forms and species. This rotting, mulchy ruined world was in some ways the cradle of humankind, for it was from this point forth that mammals came to dominate in the era called the Cenozoic: the era of new life ...

Mammals proliferated from a few small, simple, generalised forms into a diverse collection of terrestrial, marine, and flying species. However, the story of the Cenozoic is just as much about the birds (sole survivors from the dinosaur clade), the flowering plants and their relationship with insects, and the age of open grasslands.

After a cool start the world began to get warmer, and within ten million years palm trees grew in Alaska and Spitzbergen Island in the Arctic; crocodiles swam off the coasts of Greenland. The world was filled with rodent-like mammals, medium-sized mammals scavenging in forests, and larger herbivorous and carnivorous mammals hunting other mammals, birds and reptiles. There was an exponential increase in mammalian body size, leading to an almost thousand-fold increase in the weight of the largest herbivores. Atmospheric oxygen levels almost doubled over this period

Over the following ten to fifteen million years the Earth became progressively cooler again as carbon dioxide was sequestered beneath the oceans and South America fully detached from Antarctica, bringing cool, deep Antarctic water to the surface. The grasses spread further, effectively dominating a large portion of the world, diminishing forests in the process. Geological change was ongoing and rapid: the Himalayas, thrusting skywards, formed an immense barrier that shifted continental weather patterns. The Great African Rift Valley opened up, spawning a new chain of volcanic mountains in east Africa. Sea levels dropped, exposing submerged coastal plains, re-establishing a land connection between Africa and Eurasia along the eastern Mediterranean Sea, providing a migration route for primates and other animals between these continents. Tropical forests were replaced by sparse woodland and grasslands. It was into this landscape that the emerging apes spread.

By 14 million years ago, the group of apes that included our ancestors was adapting to life on the edges of the expanding savannahs in Southern Europe during a relatively short global heat wave. Then less hospitable cooler conditions in the northern hemisphere once again caused many primate species to become extinct while others survived by migrating south into Africa and South Asia. About five million years ago the African apes diverged into two lines – one that led to gorillas and another to humans and chimpanzees. Then, four million years ago, a further divergence occurred which separated the ancestors of modern chimpanzees and bonobos from the first human-like primates ... our direct ancestors.

Fossilised footprints at Laetoli in modern-day Tanzania show early hominins walking upright, leisurely, together, one following in another's footprints three million years ago. The lush rift valleys of east Africa became the cradle of modern man, as our antecedents continued to diversify, their brains evolving and growing.

It's not known when a consciousness that we would recognise as our own emerged from the welter of instinct, perception, emotion and motivation that all animals share, but within a million years cranial capacity doubled and less than 100,000 years ago early humans began their global diaspora, understanding, conceiving and making the world in their own image ... shaping the world to their own ends.

Primates

The evolutionary history of the primates can be traced back 65 million years, not long after the extinction of the dinosaurs. Their subsequent history has been marked by a prolonged and dramatic increase in head size and cranial capacity.

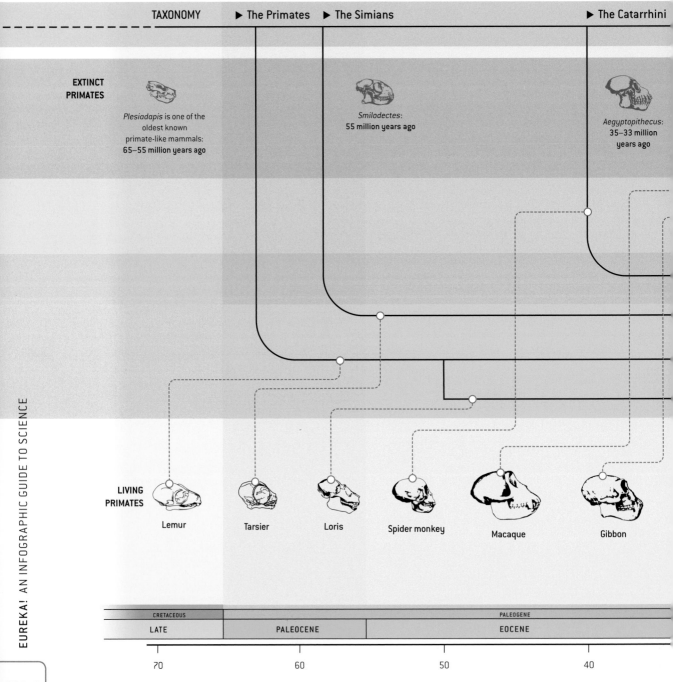

TAXONOMY ▶ The Primates ▶ The Simians ▶ The Catarrhini

EXTINCT PRIMATES

Plesiadapis is one of the oldest known primate-like mammals: 65–55 million years ago

Smilodectes: 55 million years ago

Aegyptopithecus: 35–33 million years ago

LIVING PRIMATES

Lemur

Tarsier

Loris

Spider monkey

Macaque

Gibbon

CRETACEOUS		PALEOGENE	
LATE	PALEOCENE	EOCENE	

70   60   50   40

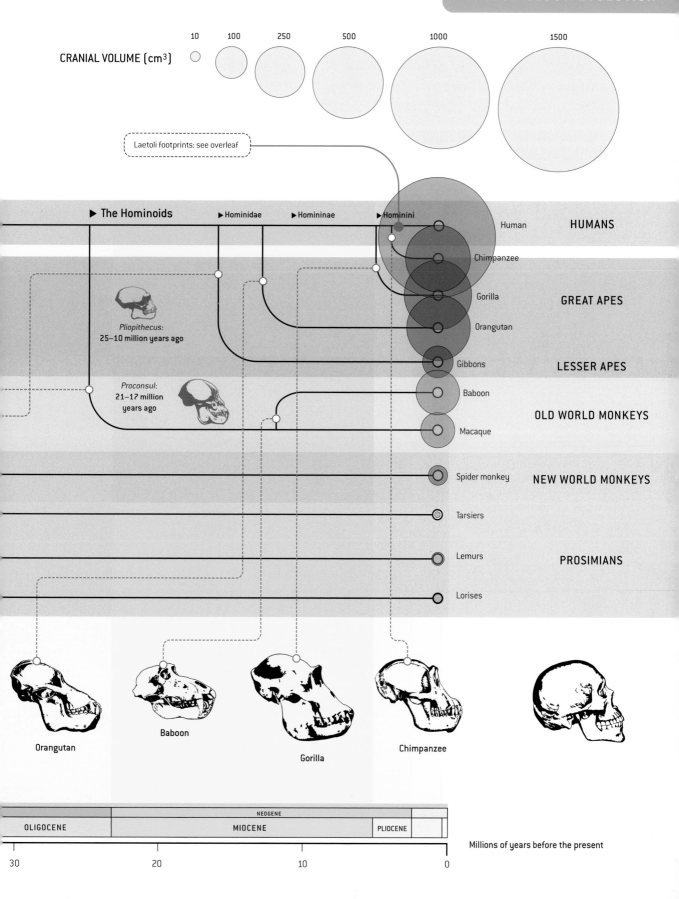

CRANIAL VOLUME (cm³)

10 100 250 500 1000 1500

Laetoli footprints: see overleaf

▶ The Hominoids   ▶ Hominidae   ▶ Homininae   ▶ Hominini

Pliopithecus:
25–10 million years ago

Proconsul:
21–17 million years ago

Human   HUMANS

Chimpanzee

Gorilla   GREAT APES

Orangutan

Gibbons   LESSER APES

Baboon   OLD WORLD MONKEYS

Macaque

Spider monkey   NEW WORLD MONKEYS

Tarsiers

Lemurs   PROSIMIANS

Lorises

Orangutan

Baboon

Gorilla

Chimpanzee

	NEOGENE		
OLIGOCENE	MIOCENE	PLIOCENE	

Millions of years before the present

30   20   10   0

Early humans

3.7 million years ago, three early humans walked through wet volcanic ash, near Laetoli, in modern-day Tanzania. Subsequent eruptions covered and preserved these oldest known footprints of early humans – probably *Australopithecus afarensis*.

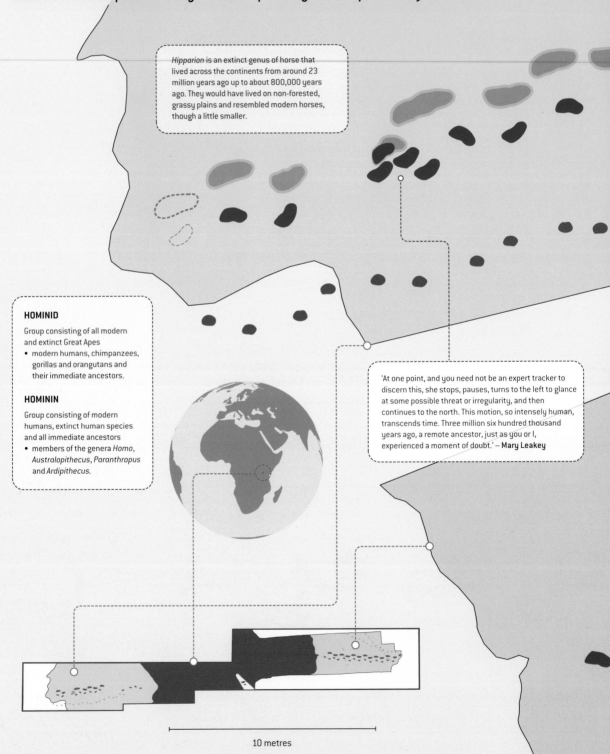

Hipparion is an extinct genus of horse that lived across the continents from around 23 million years ago up to about 800,000 years ago. They would have lived on non-forested, grassy plains and resembled modern horses, though a little smaller.

HOMINID

Group consisting of all modern and extinct Great Apes
- modern humans, chimpanzees, gorillas and orangutans and their immediate ancestors.

HOMININ

Group consisting of modern humans, extinct human species and all immediate ancestors
- members of the genera *Homo*, *Australopithecus*, *Paranthropus* and *Ardipithecus*.

'At one point, and you need not be an expert tracker to discern this, she stops, pauses, turns to the left to glance at some possible threat or irregularity, and then continues to the north. This motion, so intensely human, transcends time. Three million six hundred thousand years ago, a remote ancestor, just as you or I, experienced a moment of doubt.' – **Mary Leakey**

10 metres

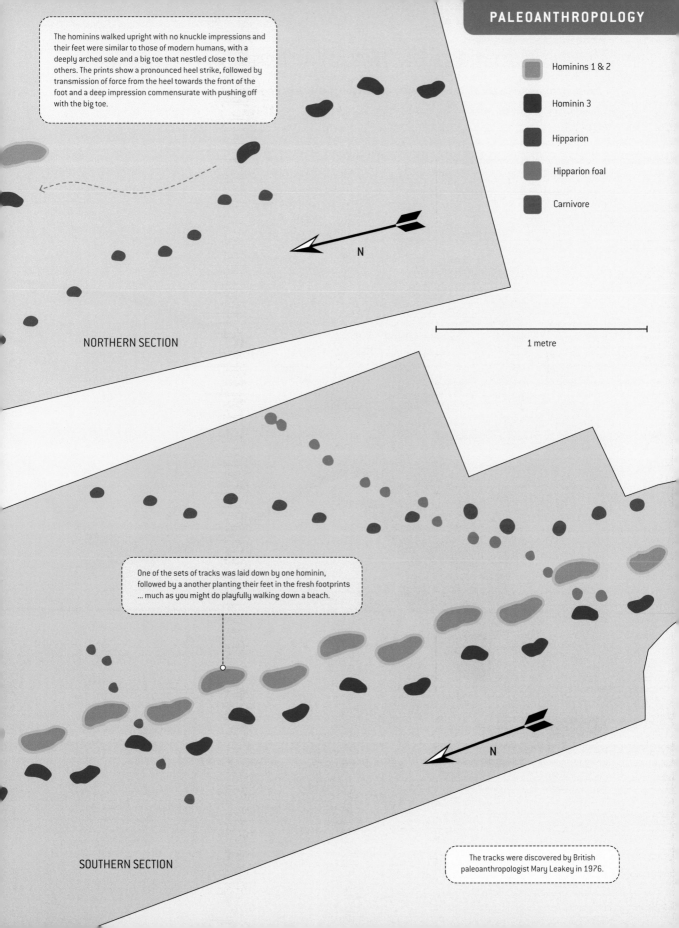

Hominins 1 & 2

Hominin 3

Hipparion

Hipparion foal

Carnivore

The hominins walked upright with no knuckle impressions and their feet were similar to those of modern humans, with a deeply arched sole and a big toe that nestled close to the others. The prints show a pronounced heel strike, followed by transmission of force from the heel towards the front of the foot and a deep impression commensurate with pushing off with the big toe.

N

1 metre

NORTHERN SECTION

One of the sets of tracks was laid down by one hominin, followed by a another planting their feet in the fresh footprints ... much as you might do playfully walking down a beach.

N

SOUTHERN SECTION

The tracks were discovered by British paleoanthropologist Mary Leakey in 1976.

The human diaspora: *Homo sapiens* on the move

The migration of modern humans out of Africa was preceded by earlier
waves of global movement by our earlier ancestors.

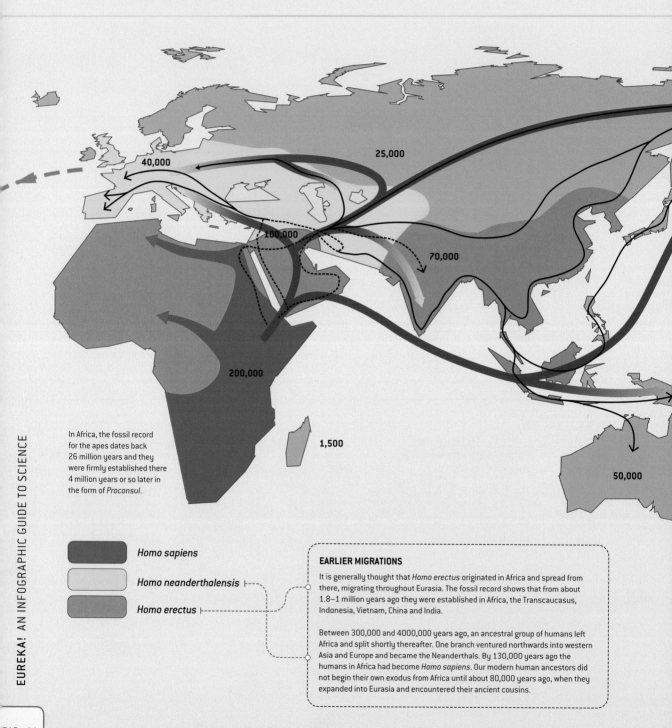

40,000

25,000

100,000

70,000

200,000

1,500

50,000

In Africa, the fossil record
for the apes dates back
26 million years and they
were firmly established there
4 million years or so later in
the form of *Proconsul*.

Homo sapiens

Homo neanderthalensis

Homo erectus

EARLIER MIGRATIONS

It is generally thought that *Homo erectus* originated in Africa and spread from
there, migrating throughout Eurasia. The fossil record shows that from about
1.8–1 million years ago they were established in Africa, the Transcaucasus,
Indonesia, Vietnam, China and India.

Between 300,000 and 4000,000 years ago, an ancestral group of humans left
Africa and split shortly thereafter. One branch ventured northwards into western
Asia and Europe and became the Neanderthals. By 130,000 years ago the
humans in Africa had become *Homo sapiens*. Our modern human ancestors did
not begin their own exodus from Africa until about 80,000 years ago, when they
expanded into Eurasia and encountered their ancient cousins.

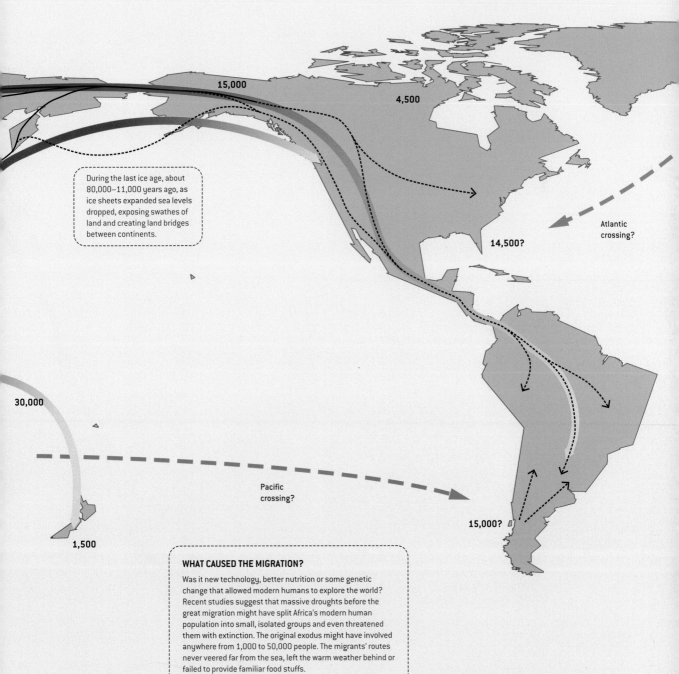

HOMO SAPIENS

→ FLOW OF GENES (mitochondrial DNA)

—— MIGRATION ROUTES

------ POSSIBLE MIGRATION ROUTES

15,000 DATE OF HUMAN SETTLEMENT (years before the present)

Plotting the genetic tree of mitochondrial DNA on a world map has produced an approximate idea of the routes taken by modern humans in their expansion. The forks in the tree corresponded to geographical separations of populations. The sequence of branches corresponds to that of the geographical separations and the position and length of the branches to the time at which the splits occurred.

15,000

4,500

During the last ice age, about 80,000–11,000 years ago, as ice sheets expanded sea levels dropped, exposing swathes of land and creating land bridges between continents.

Atlantic crossing?

14,500?

30,000

Pacific crossing?

15,000?

1,500

WHAT CAUSED THE MIGRATION?

Was it new technology, better nutrition or some genetic change that allowed modern humans to explore the world? Recent studies suggest that massive droughts before the great migration might have split Africa's modern human population into small, isolated groups and even threatened them with extinction. The original exodus might have involved anywhere from 1,000 to 50,000 people. The migrants' routes never veered far from the sea, left the warm weather behind or failed to provide familiar food stuffs.

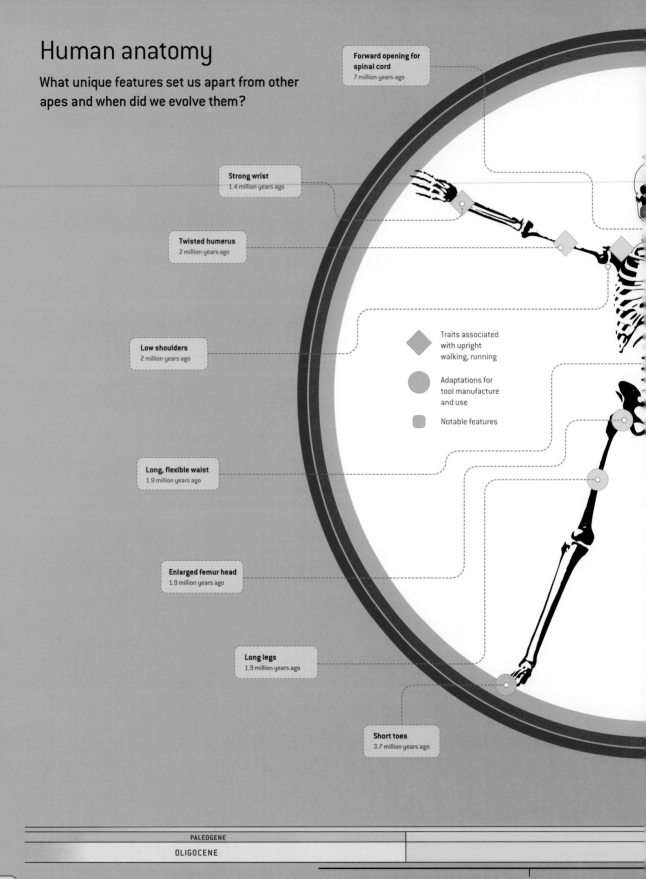

Human anatomy

What unique features set us apart from other apes and when did we evolve them?

Forward opening for spinal cord
7 million years ago

Strong wrist
1.4 million years ago

Twisted humerus
2 million years ago

Low shoulders
2 million years ago

Traits associated with upright walking, running

Adaptations for tool manufacture and use

Notable features

Long, flexible waist
1.9 million years ago

Enlarged femur head
1.9 million years ago

Long legs
1.9 million years ago

Short toes
3.7 million years ago

PALEOGENE

OLIGOCENE

Millions of years before the present   20

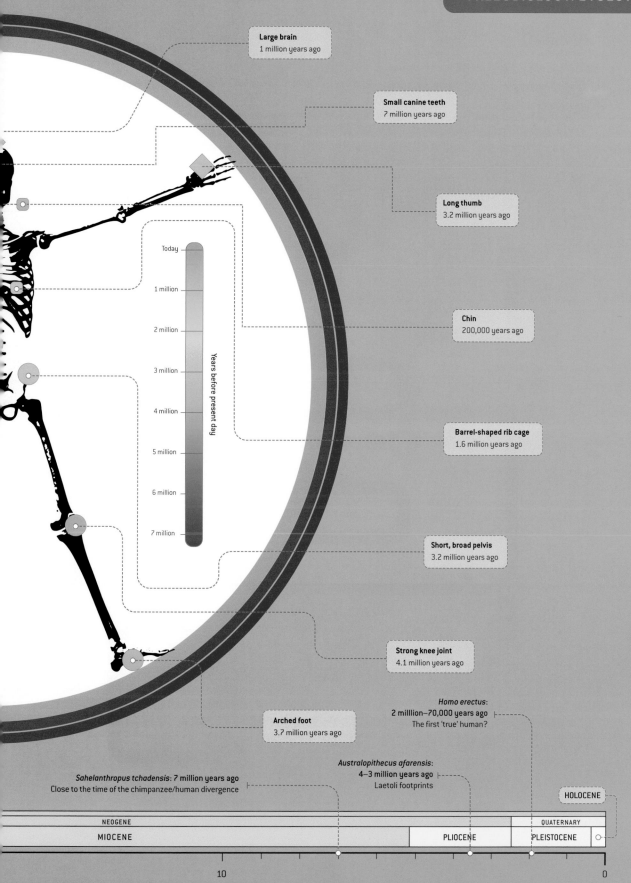

Large brain
1 million years ago

Small canine teeth
7 million years ago

Long thumb
3.2 million years ago

Chin
200,000 years ago

Barrel-shaped rib cage
1.6 million years ago

Short, broad pelvis
3.2 million years ago

Today

1 million

2 million

3 million

4 million

5 million

6 million

7 million

Years before present day

Strong knee joint
4.1 million years ago

Homo erectus:
2 milllion–70,000 years ago
The first 'true' human?

Arched foot
3.7 million years ago

Australopithecus afarensis:
4–3 million years ago
Laetoli footprints

Sahelanthropus tchadensis: **7 million years ago**
Close to the time of the chimpanzee/human divergence

HOLOCENE

NEOGENE

QUATERNARY

MIOCENE

PLIOCENE

PLEISTOCENE

10

0

Digestion

The human digestive tract is an average of five metres long with a surface area of about 32 square metres.

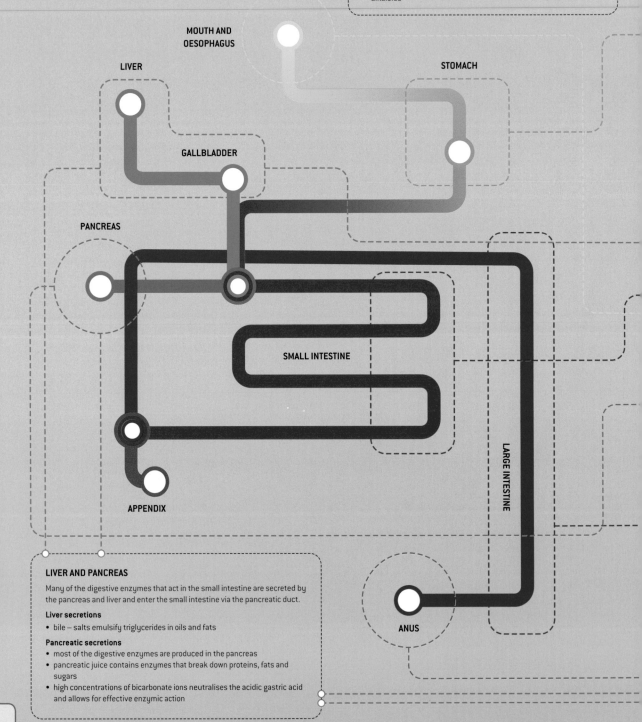

MOUTH

Secretion
- mucus
- wide range of ions that buffer the pH of the mouth
- salivary amylase, an enzyme which breaks down starch to sugars such as maltose and dextrin. About 30 per cent of starch is digested in the mouth
- salivary lipase an enzyme that begins fat digestion
- antibacterial compounds and enzymes

Absorption
- glucose, alcohol, certain water-soluble drugs; low molecular weight alkaloids

MOUTH AND OESOPHAGUS

LIVER

STOMACH

GALLBLADDER

PANCREAS

SMALL INTESTINE

LARGE INTESTINE

APPENDIX

ANUS

LIVER AND PANCREAS

Many of the digestive enzymes that act in the small intestine are secreted by the pancreas and liver and enter the small intestine via the pancreatic duct.

Liver secretions
- bile – salts emulsify triglycerides in oils and fats

Pancreatic secretions
- most of the digestive enzymes are produced in the pancreas
- pancreatic juice contains enzymes that break down proteins, fats and sugars
- high concentrations of bicarbonate ions neutralises the acidic gastric acid and allows for effective enzymic action

pH

ABSORPTION AND EXCRETION

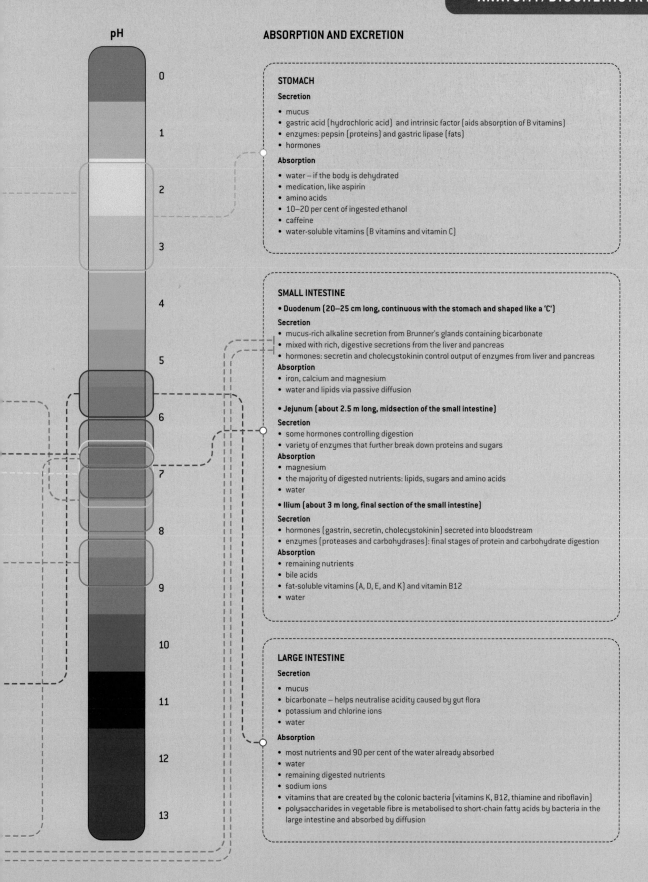

0

1

2

3

4

5

6

7

8

9

10

11

12

13

STOMACH

Secretion

- mucus
- gastric acid (hydrochloric acid) and intrinsic factor (aids absorption of B vitamins)
- enzymes: pepsin (proteins) and gastric lipase (fats)
- hormones

Absorption

- water – if the body is dehydrated
- medication, like aspirin
- amino acids
- 10–20 per cent of ingested ethanol
- caffeine
- water-soluble vitamins (B vitamins and vitamin C)

SMALL INTESTINE

- **Duodenum (20–25 cm long, continuous with the stomach and shaped like a 'C')**

Secretion

- mucus-rich alkaline secretion from Brunner's glands containing bicarbonate
- mixed with rich, digestive secretions from the liver and pancreas
- hormones: secretin and cholecystokinin control output of enzymes from liver and pancreas

Absorption

- iron, calcium and magnesium
- water and lipids via passive diffusion

- **Jejunum (about 2.5 m long, midsection of the small intestine)**

Secretion

- some hormones controlling digestion
- variety of enzymes that further break down proteins and sugars

Absorption

- magnesium
- the majority of digested nutrients: lipids, sugars and amino acids
- water

- **Ilium (about 3 m long, final section of the small intestine)**

Secretion

- hormones (gastrin, secretin, cholecystokinin) secreted into bloodstream
- enzymes (proteases and carbohydrases): final stages of protein and carbohydrate digestion

Absorption

- remaining nutrients
- bile acids
- fat-soluble vitamins (A, D, E, and K) and vitamin B12
- water

LARGE INTESTINE

Secretion

- mucus
- bicarbonate – helps neutralise acidity caused by gut flora
- potassium and chlorine ions
- water

Absorption

- most nutrients and 90 per cent of the water already absorbed
- water
- remaining digested nutrients
- sodium ions
- vitamins that are created by the colonic bacteria (vitamins K, B12, thiamine and riboflavin)
- polysaccharides in vegetable fibre is metabolised to short-chain fatty acids by bacteria in the large intestine and absorbed by diffusion

The circulatory system

With a combined length of about 100,000 km, your blood vessels could circumnavigate the globe two and a half times; and your heart might beat over 3 billion times before it's done.

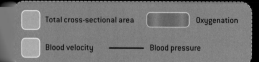

Total cross-sectional area • Oxygenation

Blood velocity — Blood pressure

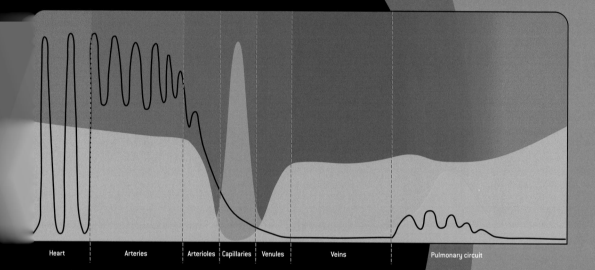

Heart · Arteries · Arterioles · Capillaries · Venules · Veins · Pulmonary circuit

SIZE OF BLOOD VESSELS MAGNIFIED 100 TIMES

ARTERIOLE
- Average diameter: 0.03 mm
- Average wall thickness: 0.006 mm

VENULE
- Average diameter: 0.02 mm
- Average wall thickness: 0.001 mm

CAPILLARY
- Average diameter: 0.008 mm
- Average wall thickness: 0.0005 mm

ARTERY
- Average diameter: 4 mm
- Average wall thickness: 1 mm

VEIN
- Average diameter: 5 mm
- Average wall thickness: 0.5 mm

PROPORTIONATE VOLUME OF THE CIRCULATORY SYSTEM

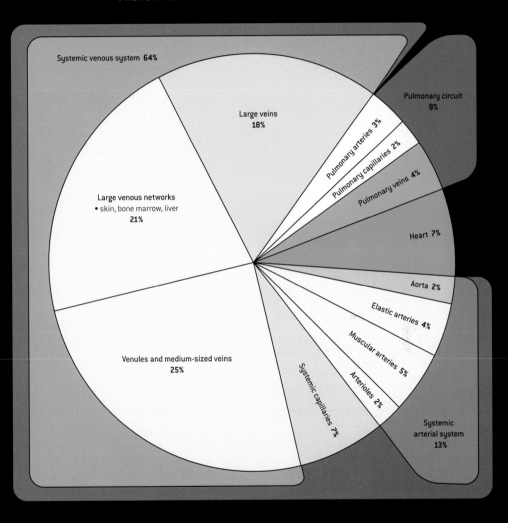

Systemic venous system **64%**

Large veins
18%

Pulmonary circuit
9%

Pulmonary arteries **3%**

Pulmonary capillaries **2%**

Pulmonary veins **4%**

Heart **7%**

Large venous networks
• skin, bone marrow, liver
21%

Aorta **2%**

Elastic arteries **4%**

Muscular arteries **5%**

Arterioles **2%**

Venules and medium-sized veins
25%

Systemic capillaries **7%**

Systemic
arterial system
13%

Hormones

The endocrine system is an information signal system like the nervous system, but with slower effects and much more prolonged response. But the two systems are intimately linked and form an integrated command and control system.

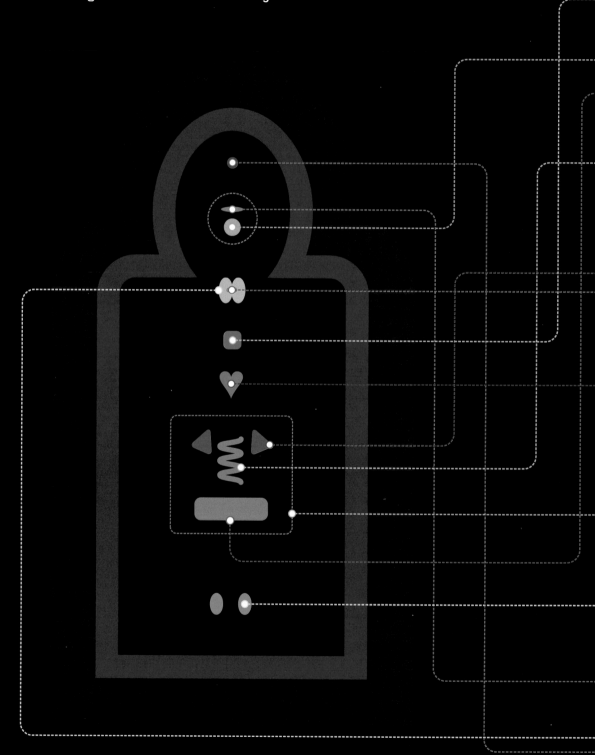

...he property of a system in which a huge range of interconnected and complex processes within the human body, that tend to change due to internal ...r environmental forces, are actively regulated to maintain balance, to remain very nearly constant, or return the system to functioning within a ...ormal range. The brain directs this process via the hypothalamus, the autonomic nervous system and the **endocrine system** – the collection of ...ands within a human that secrete hormones directly into the circulatory system to be carried towards distant target organs.

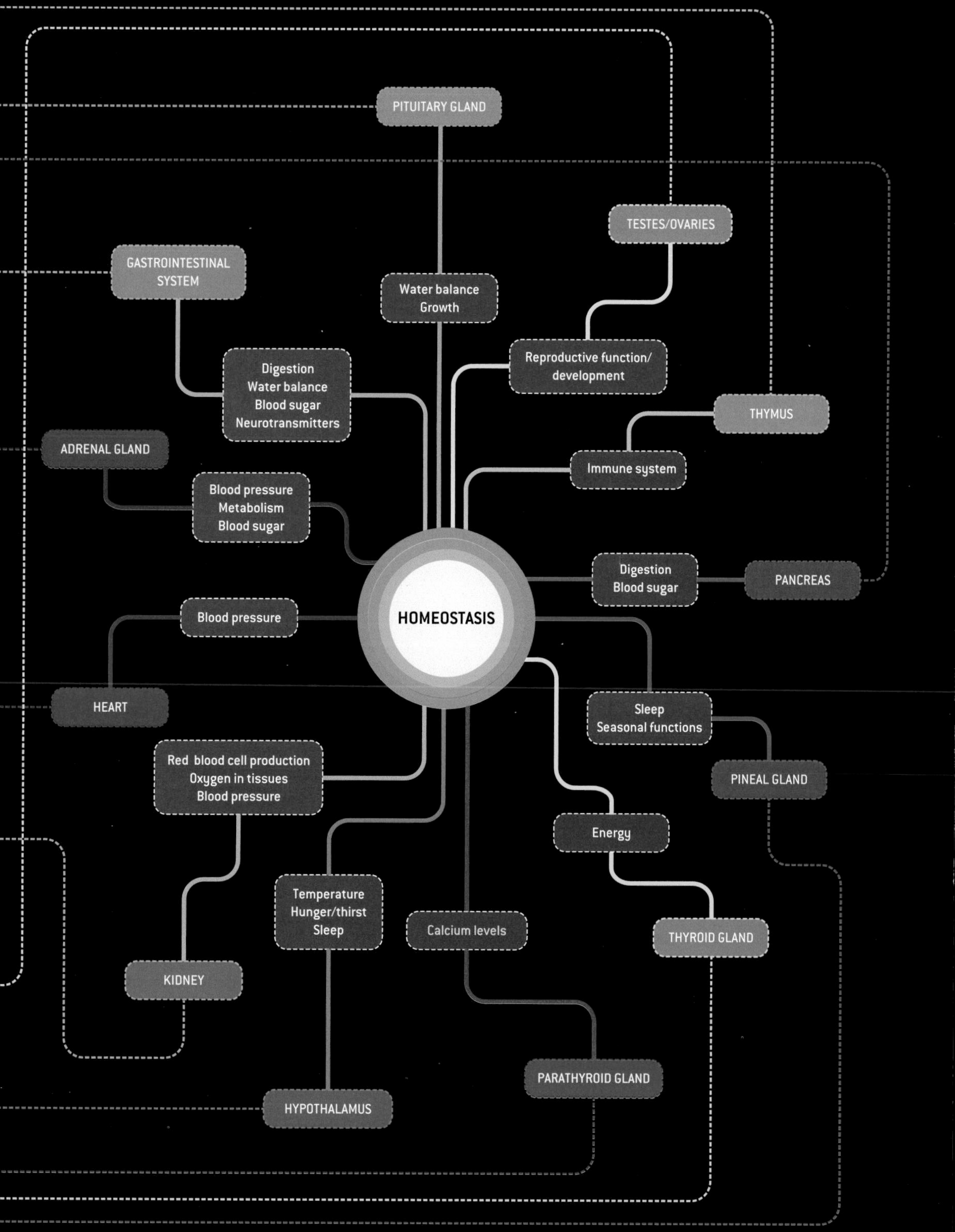

PITUITARY GLAND

TESTES/OVARIES

GASTROINTESTINAL
SYSTEM

Water balance
Growth

Digestion
Water balance
Blood sugar
Neurotransmitters

Reproductive function/
development

THYMUS

ADRENAL GLAND

Blood pressure
Metabolism
Blood sugar

Immune system

Digestion
Blood sugar

PANCREAS

Blood pressure

HOMEOSTASIS

HEART

Sleep
Seasonal functions

Red blood cell production
Oxygen in tissues
Blood pressure

PINEAL GLAND

Energy

Temperature
Hunger/thirst
Sleep

Calcium levels

THYROID GLAND

KIDNEY

PARATHYROID GLAND

HYPOTHALAMUS

Language evolution

Estimates of the number of living languages in the world vary between 5,000 and 7,000. Language originated when early hominins started gradually changing their primate communication systems, acquiring the ability to form a theory of other minds, empathy and a shared intentionality.

LIFESPANS OF WRITTEN LANGUAGES

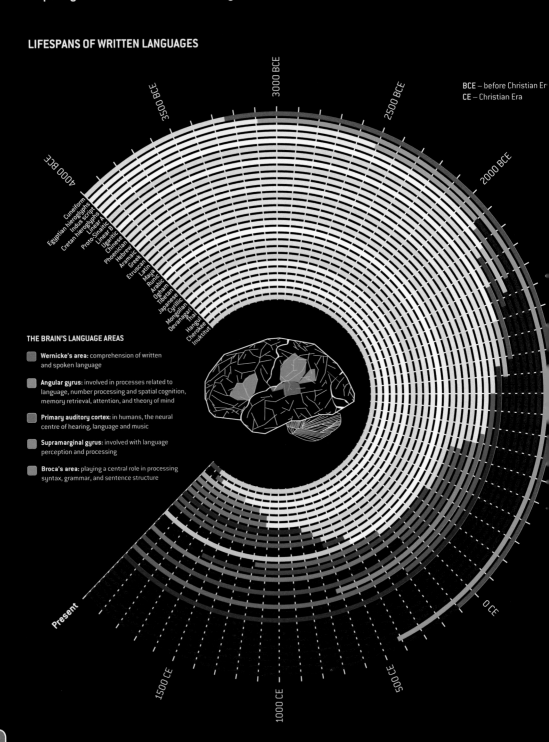

BCE – before Christian Er
CE – Christian Era

3500 BCE
4000 BCE
3000 BCE
2500 BCE
2000 BCE

Cuneiform
Egyptian hieroglyphs
Indus script
Cretan hieroglyphs
Linear A
Proto-Sinaitic
Linear B
Ugaritic
Chinese
Hebrew
Phoenician
Aramaic
Greek
Etruscan
Latin
Maya
Runic
Arabic
Ogham
Tibetan
Japanese
Cyrillic
Mongolian
Devanagari
Thai
Hangul
Cherokee
Inuktitut

THE BRAIN'S LANGUAGE AREAS

- **Wernicke's area:** comprehension of written and spoken language

- **Angular gyrus:** involved in processes related to language, number processing and spatial cognition, memory retrieval, attention, and theory of mind

- **Primary auditory cortex:** in humans, the neural centre of hearing, language and music

- **Supramarginal gyrus:** involved with language perception and processing

- **Broca's area:** playing a central role in processing syntax, grammar, and sentence structure

Present
1500 CE
1000 CE
500 CE
0 CE

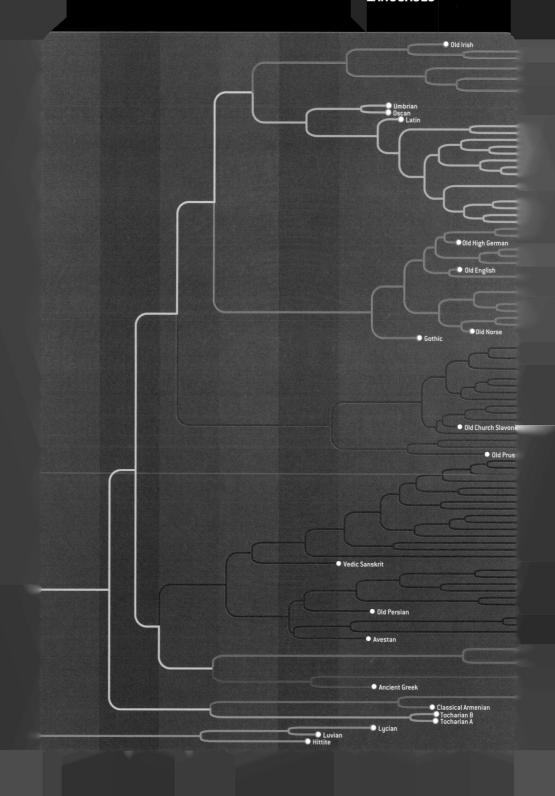

Old Irish

Umbrian
Oscan
Latin

Old High German

Old English

Old Norse

Gothic

Old Church Slavoni

Old Prus

Vedic Sanskrit

Old Persian

Avestan

Ancient Greek

Classical Armenian
Tocharian B
Tocharian A

Lycian
Luvian
Hittite

Brain evolution: size and complexity

Human (23,000 million)

Number of neurons in the cerebral cortex ...

African elephant (11,000 million)

Chimpanzee (6,200 millon)

Cat (300 million)

Dog (160 million)

Pilot whale (3,000 million)

Fin whale (1,500 million)

Mouse (4 million)

Rat (21 million)

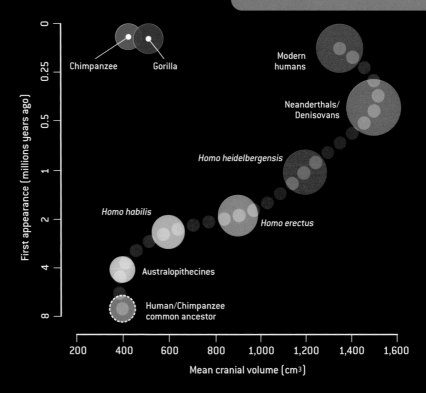

First appearance (millions years ago)

Chimpanzee

Gorilla

Modern humans

Neanderthals/ Denisovans

Homo heidelbergensis

Homo habilis

Homo erectus

Australopithecines

Human/Chimpanzee common ancestor

Mean cranial volume (cm³)

The **Human Connectome Project** is an international research project that aims to build a 'network map' of the human brain's connectivity, focusing principally on the brain's 'wiring' – the white, myelin-rich tracts – using functional magnetic resonance imaging. The hope is that this will shed light on the anatomical and functional connectivity within the human brain and produce a body of data that will aid research into brain disorders. To the right is a representation of brain connectivity obtained using magnetic resonance imaging. Colours of the outermost ring map to regional areas of the cortex. The innermost rings represent average regional volume; surface area; thickness; curvature; and degree of specific connectedness, respectively. The development of **strong Artificial Intelligence** may also depend on mirroring the living connections of the human brain in the computers of the future.

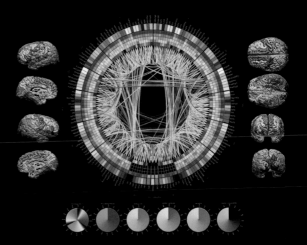

Total length of myelinated fibres in central nervous system of a typical 20-year-old human

male	176,000 km
female	149,000 km
Distance to the moon from Earth	406,000 km

Perception

Distinct, specific and yet profoundly integrated, our perceptual systems provide us with a rich simulacrum of the Universe.

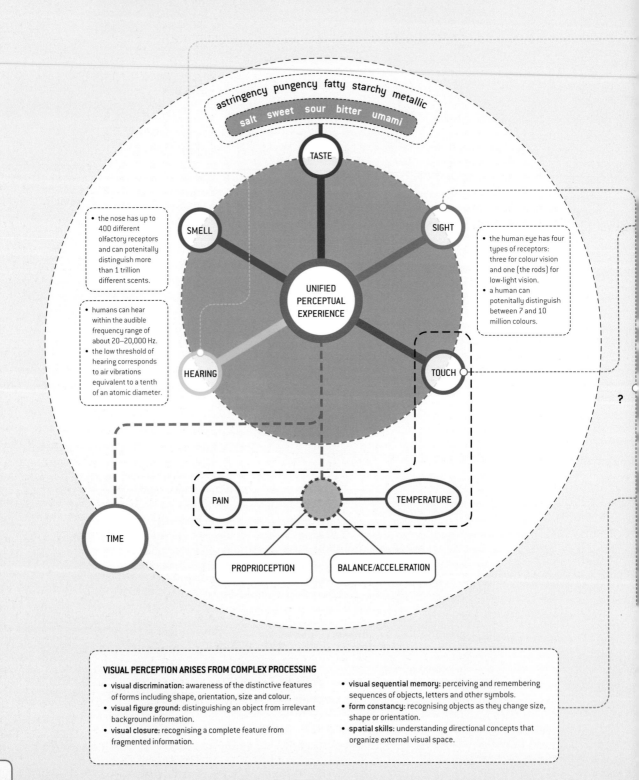

astringency pungency fatty starchy metallic

salt sweet sour bitter umami

TASTE

SIGHT

SMELL

- the nose has up to 400 different olfactory receptors and can potenitally distinguish more than 1 trillion different scents.

- humans can hear within the audible frequency range of about 20–20,000 Hz.
- the low threshold of hearing corresponds to air vibrations equivalent to a tenth of an atomic diameter.

UNIFIED PERCEPTUAL EXPERIENCE

- the human eye has four types of receptors: three for colour vision and one (the rods) for low-light vision.
- a human can potenitally distinguish between 7 and 10 million colours.

HEARING

TOUCH

?

PAIN

TEMPERATURE

TIME

PROPRIOCEPTION

BALANCE/ACCELERATION

VISUAL PERCEPTION ARISES FROM COMPLEX PROCESSING

- **visual discrimination**: awareness of the distinctive features of forms including shape, orientation, size and colour.
- **visual figure ground**: distinguishing an object from irrelevant background information.
- **visual closure**: recognising a complete feature from fragmented information.

- **visual sequential memory**: perceiving and remembering sequences of objects, letters and other symbols.
- **form constancy**: recognising objects as they change size, shape or orientation.
- **spatial skills**: understanding directional concepts that organize external visual space.

SENSES WORKING OVERTIME

ECHOLOCATION

Echolocating animals emit calls out to the environment and listen to the echoes of those calls. They use these echoes to locate and identify the objects. Echolocation is used for navigation, for foraging and hunting. Animals that use echolocation are cheifly mammals – bats, dolphins, porpoises and toothed whales – but also a few bird species that are nocturnal or cave-dwelling, though their sense is crude by comparison with mammalian echolocators.

Bat echolocation frequencies range from as low as 11,000 Hz to as high as 200,000 kHz. For comparison, human hearing range is about 20–20,000 Hz.

Blind humans can learn to find their way using echolocating clicks produced by a mechanical device or by mouth.

ELECTRORECEPTION

Electroreception is the ability to perceive electrical fields: it was thought that electroreception was only present in vertebrates, but recent research has demonstrated that bees can percieve the strength and pattern of electrostatic charge on flower petals. Electroreception is found in lampreys, cartilaginous fishes (sharks and rays), catfish, lungfish, coelacanths, sturgeon and paddlefish – but also in a few mammals, notably the montreme marsupials, the duck billed platypus and echidna; and one species of cetacean, the Guiana dolphin. The electroreceptor organs in all these groups are derived embryologically from a mechanoreceptor system.

MAGNETORECEPTION

Magnetoreception is a sense which allows an organism, the most studied and understood example being migratory birds, to detect the pervasive magnetic field to construct detailed internal maps of local variations and features, which are then used to ascertain the animal's direction, altitude and location for the purpose of navigation.

This magnetic sense in birds has surprising properties: magnetoreception is located in the eye and requires the presence of light to operate – a bird can only sense a magnetic field if certain wavelengths of bluish light are available. The exact mechanism of this sense remains uncertain, though it may involve a special protein called **cryptochrome**. The avian compass is also an **inclination-only compass**, meaning that it can sense changes in the inclination of magnetic field lines – their angle relative to the horizontal, but is not sensitive to the polarity of the field lines. Under natural conditions, birds are sensitive to a very narrow band of magnetic field strengths around the geomagnetic field strength, but can orient at higher or lower field strengths when trained under experimental conditions.

POLARISED LIGHT DETECTION

Visual pigments, the molecules in photoreceptors that initiate the process of vision, are inherently **dichroic** – meaning they absorb light differently according to its axis of polarisation. Humans can subtly perceive the linear polarisation of light but this ability is thought to be a byproduct of the the structure of the lens or central region of the retina and has no biological function. True perceptual sensitivity to polarised light has evolved many times independently and is found in a wide range of animals, including many orders of insects, spiders, crustaceans, cephalopods, fish, amphibians, reptiles, and birds. This ensitivity is generally associated with behavioral tasks like orientation or navigation, but it can be incorporated into a high-level visual perception akin to colour vision, allowing the animal to percieve a viewed scene as regions that differ in their polarisation. This contrast enhancement, camouflage breaking, object recognition and signal detection – particularly in deeper-water environments.

Taste

This flavour wheel developed by the Speciality Coffee Association of America demonstrates the enormous complexity of perceptual experience that can be contained in one mouthful of good coffee ...

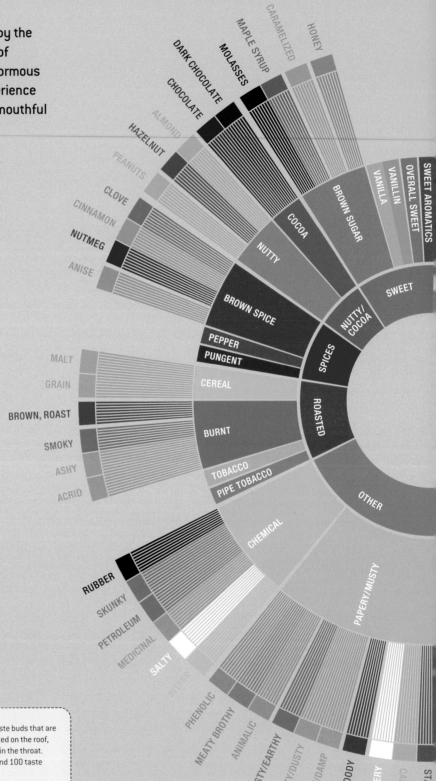

There are between 2,000 and 5,000 taste buds that are located on the tongue. Others are located on the roof, sides and back of the mouth and even in the throat. Each taste bud contains between 50 and 100 taste receptor cells.

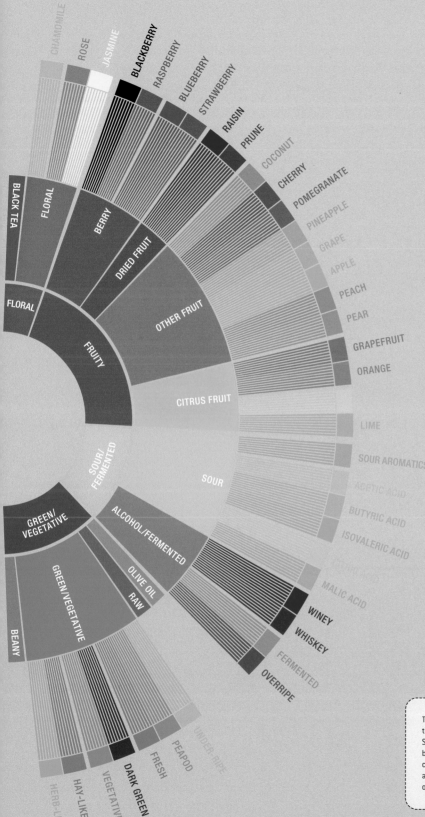

FURTHER TASTE DIMENSIONS

- coolness
- pungency/spiciness/heat
- astringency
- numbness
- metallicness
- calcium (chalkiness)
- fattiness/oiliness
- temperature

Taste buds resolve different tastes through detecting the interaction between different molecules or ions. Sweet, umami, and bitter tastes are triggered by the binding of molecules specific receptor proteins in the cell membranes of taste buds. Saltiness and sourness are perceived when taste buds encounter alkali metal or hydrogen ions, respectively.

CHAMOMILE
ROSE
JASMINE
BLACKBERRY
RASPBERRY
BLUEBERRY
STRAWBERRY
RAISIN
PRUNE
COCONUT
CHERRY
POMEGRANATE
PINEAPPLE
GRAPE
APPLE
PEACH
PEAR
GRAPEFRUIT
ORANGE
LIME
SOUR AROMATICS
ACETIC ACID
BUTYRIC ACID
ISOVALERIC ACID
CITRIC ACID
MALIC ACID
WINEY
WHISKEY
FERMENTED
OVERRIPE
UNDER-RIPE
PEAPOD
FRESH
VEGETATIVE
DARK GREEN
HAY-LIKE
HERB-LIKE

BLACK TEA
FLORAL
BERRY
DRIED FRUIT
OTHER FRUIT
CITRUS FRUIT
FLORAL
FRUITY
SOUR
SOUR/FERMENTED
GREEN/VEGETATIVE
ALCOHOL/FERMENTED
OLIVE OIL
RAW
GREEN/VEGETATIVE
BEANY

Neurotransmission

Communication between neurons is accomplished by the movement of chemicals, or neurotransmitters, across a synapse. Neurons receive input information from other neurons across synapses, process that information and then send the information as output to other neurons or muscles through synapses.

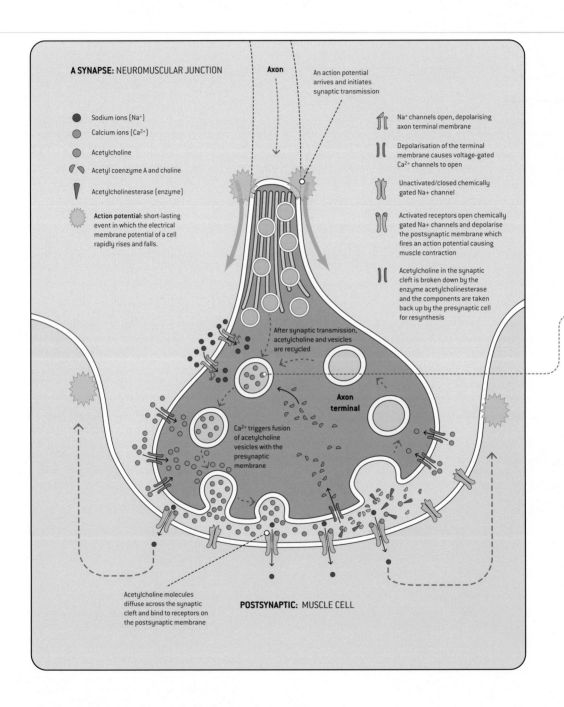

A SYNAPSE: NEUROMUSCULAR JUNCTION

Axon

An action potential arrives and initiates synaptic transmission

- Sodium ions (Na^+)
- Calcium ions (Ca^{2+})
- Acetylcholine
- Acetyl coenzyme A and choline
- Acetylcholinesterase (enzyme)

Action potential: short-lasting event in which the electrical membrane potential of a cell rapidly rises and falls.

Na^+ channels open, depolarising axon terminal membrane

Depolarisation of the terminal membrane causes voltage-gated Ca^{2+} channels to open

Unactivated/closed chemically gated Na+ channel

Activated receptors open chemically gated Na+ channels and depolarise the postsynaptic membrane which fires an action potential causing muscle contraction

Acetylcholine in the synaptic cleft is broken down by the enzyme acetylcholinesterase and the components are taken back up by the presynaptic cell for resynthesis

After synaptic transmission, acetylcholine and vesicles are recycled

Axon terminal

Ca^{2+} triggers fusion of acetylcholine vesicles with the presynaptic membrane

Acetylcholine molecules diffuse across the synaptic cleft and bind to receptors on the postsynaptic membrane

POSTSYNAPTIC: MUSCLE CELL

NEUROTRANSMITTERS: THEIR FUNCTION, CHEMICAL STRUCTURES AND SIZE

NORADRENALINE
Molecular mass: 169

DOPAMINE
Molecular mass: 153

SMALL MONOAMINES

Dopamine

Plays a major role in reward-motivated behaviour: most types of reward increase the level of dopamine in the brain and addictive drugs increase dopamine activity. Outside of the central nervous system dopamine plays an important role in the immune system, the kidneys and the pancreas. Dopamine is also a precursor of noradrenaline and adrenaline.

Noradrenaline

Synthesised and released by the central nervous system and also by the sympathetic division of the autonomic nervous system, the action of which mobilises the brain and body for action. Noradrenaline release is lowest during sleep, rises during wakefulness and reaches highest levels during situations of stress or danger – the so-called fight-or-flight response.

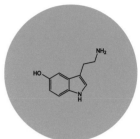

SEROTONIN
Molecular mass: 176

MONOAMINE

GABA

The main inhibitory neurotransmitter in the human central nervous system, reducing neuronal excitability throughout the nervous system. It is also responsible for the regulation of muscle tone.

Glutamate

The most abundant excitatory neurotransmitter in the nervous system; it is involved in cognitive functions such as learning and memory

Serotonin

The majority of the human body's serotonin is located in gastrointestinal tract, where it is used to regulate intestinal movements. The remainder is located in the central nervous system, where it is involved in the regulation of mood, appetite, and sleep. Serotonin plays a role in some cognitive functions, including memory and learning.

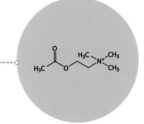

ACETYLCHOLINE
Molecular mass: 146

γ-AMINOBUTYRIC ACID (GABA)
Molecular mass: 103

GLUTAMATE
Molecular mass: 147

SMALL AMINO ACIDS

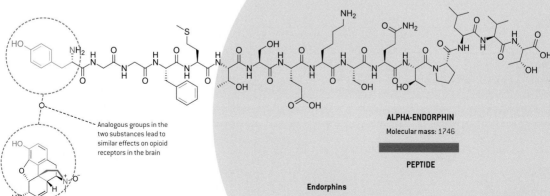

Analogous groups in the two substances lead to similar effects on opioid receptors in the brain

ALPHA-ENDORPHIN
Molecular mass: 1746

PEPTIDE

MORPHINE
Molecular mass: 285

Endorphins

Bind to opioid receptors in the brain: opiates, such as morphine and codeine, mimic the effect of these peptides. Endorphins are known to play an important role in motivation, emotion, attachment behaviour and the response to stress and pain. They are naturally produced in response to pain, but can also be triggered by activities such as vigorous exercise ... and laughter.

Emotion

Psychologists attempt to anatomise those most intense, slippery and uncognitive qualities of the human mind ... our needs and emotions.

PLUTCHIK'S WHEEL OF EMOTION

Basic emotion	Opposite
Joy	Sadness
Trust	Disgust
Fear	Anger
Surprise	Anticipation

Emotional composites
Anticipation + Joy
Joy + Trust
Trust + Fear
Fear + Surprise
Surprise + Sadness
Sadness + Disgust
Disgust + Anger
Anger + Anticipation

Derived feelings
Optimism
Love
Submission
Awe
Disapproval
Remorse
Contempt
Aggressiveness

Opposite feeling
Disapproval
Remorse
Contempt
Aggression
Optimism
Love
Submission
Awe

- **Loathing** Intensified emotion
- **JOY** Basic emotion
- **apprehension** Mild emotion
- **REMORSE** Composite feeling
- Emotional neutrality

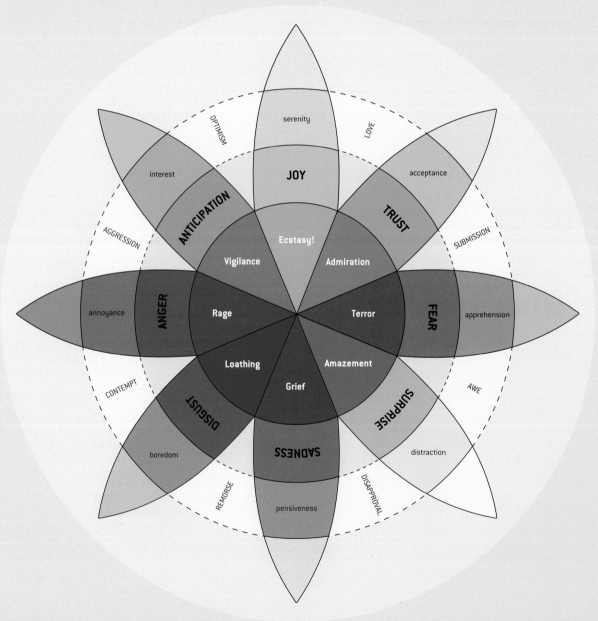

MASLOW'S PYRAMID OF HUMAN NEEDS

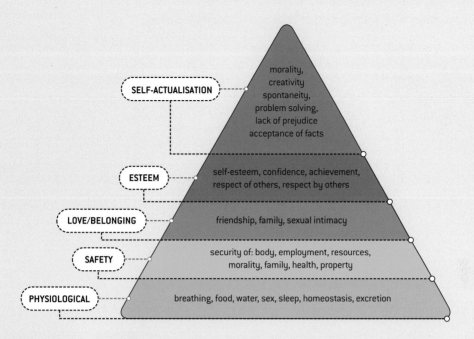

SELF-ACTUALISATION
morality,
creativity
spontaneity,
problem solving,
lack of prejudice
acceptance of facts

ESTEEM
self-esteem, confidence, achievement,
respect of others, respect by others

LOVE/BELONGING
friendship, family, sexual intimacy

SAFETY
security of: body, employment, resources,
morality, family, health, property

PHYSIOLOGICAL
breathing, food, water, sex, sleep, homeostasis, excretion

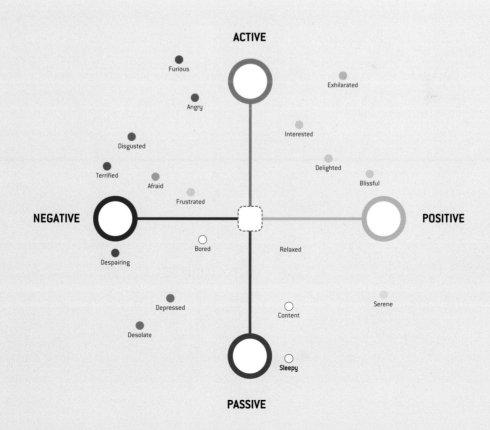

ACTIVE

Furious

Exhilarated

Angry

Interested

Disgusted

Delighted

Terrified

Blissful

Afraid

Frustrated

NEGATIVE · POSITIVE

Bored

Relaxed

Despairing

Depressed

Serene

Content

Desolate

Sleepy

PASSIVE

Learning and memory: information processing

A central tenet of cognitive science is that a complete understanding of the brain and mind cannot be attained by studying only a single level.

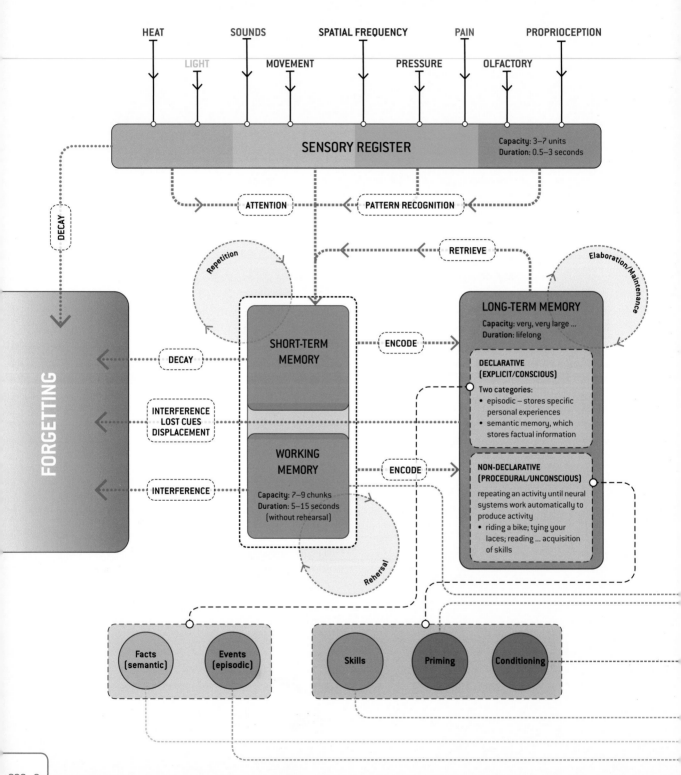

LEARNING, REMEMBERING, REHEARSAL AND FORGETTING

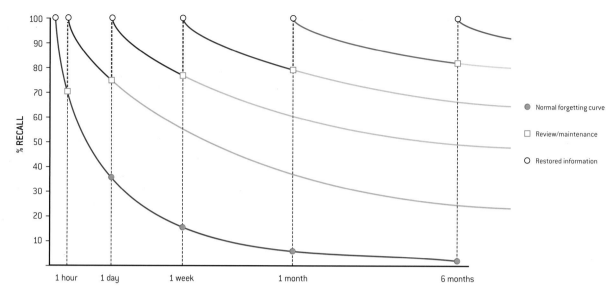

CONCIOUS LEARNING

Declarative memory will naturally decay in an exponential way over time and be lost to recall. Repeated review of the facts or episodic details at decreasing intervals restores the information and subsequent decay is slower and shallower. Eventually the information will be permanently stored, with relatively minimal decay.

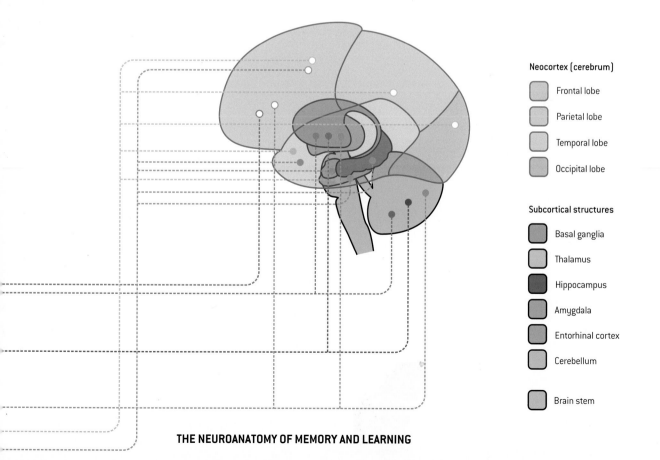

THE NEUROANATOMY OF MEMORY AND LEARNING

Computing

The exponential growth in computing power that has driven extraordinary changes in our everyday life, is founded on an explosive increase in the density and scale of circuit integration. Silicon chips lie at the heart of this revolution of human civilisation and point to a future where thinking machines might surpass human intelligence.

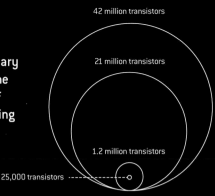

42 million transistors

21 million transistors

1.2 million transistors

25,000 transistors

size of chip mm²

5   300

180   scale of smallest components (nanometres)

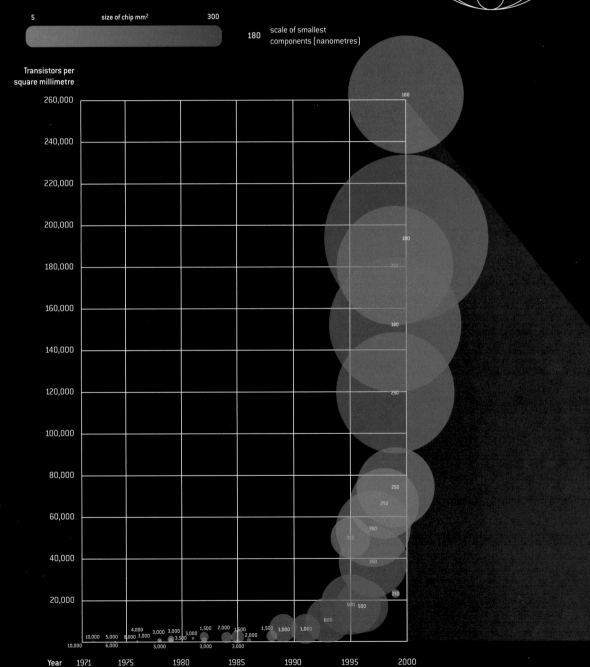

Transistors per square millimetre

260,000

240,000

220,000

200,000

180,000

160,000

140,000

120,000

100,000

80,000

60,000

40,000

20,000

180
180
250
180
250
250
250
350
350
350
350
500 500
600
10,000 5,000
8,000 3,000
3,000 3,000
3,500
4,000
1,500 2,000
1,500
2,000
1,500 1,000
1,000

10,000   6,000   5,000   3,000   3,000

Year   1971   1975   1980   1985   1990   1995   2000

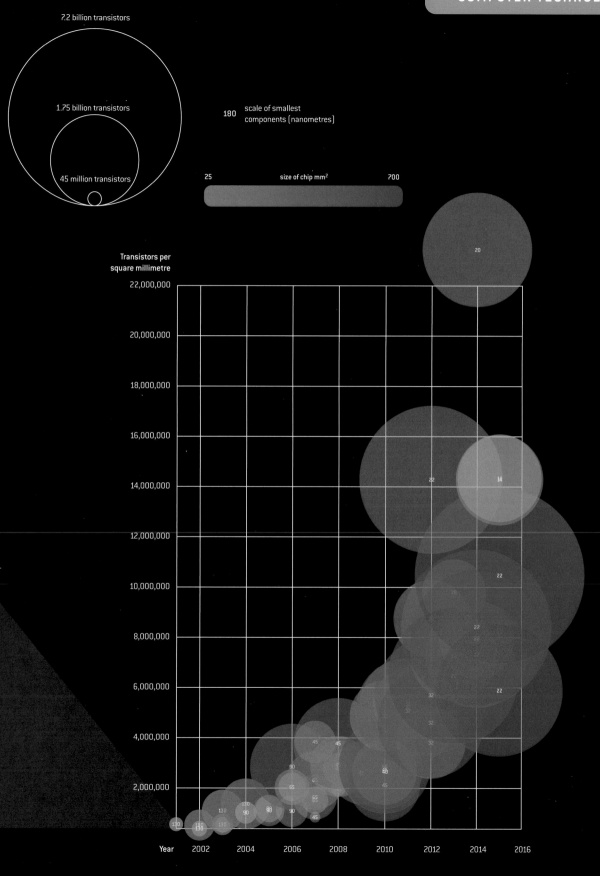

7.2 billion transistors

1.75 billion transistors

45 million transistors

180   scale of smallest
       components (nanometres)

25   size of chip mm²   700

Transistors per
square millimetre

22,000,000

20,000,000

18,000,000

16,000,000

14,000,000

12,000,000

10,000,000

8,000,000

6,000,000

4,000,000

2,000,000

Year   2002   2004   2006   2008   2010   2012   2014   2016

Artificial intelligence

With the exponential growth and development of computers over the last fifty years, it now seems that we are on the verge of creating true computer intelligence to rival or even surpass human intelligence. But can true humanness ever be digitially replicated or improved upon?

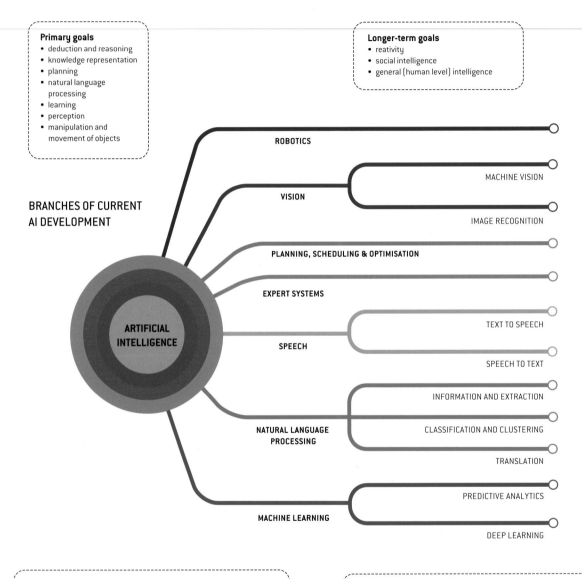

Primary goals
- deduction and reasoning
- knowledge representation
- planning
- natural language processing
- learning
- perception
- manipulation and movement of objects

Longer-term goals
- reativity
- social intelligence
- general (human level) intelligence

BRANCHES OF CURRENT AI DEVELOPMENT

ARTIFICIAL INTELLIGENCE

- ROBOTICS
- VISION
  - MACHINE VISION
  - IMAGE RECOGNITION
- PLANNING, SCHEDULING & OPTIMISATION
- EXPERT SYSTEMS
- SPEECH
  - TEXT TO SPEECH
  - SPEECH TO TEXT
- NATURAL LANGUAGE PROCESSING
  - INFORMATION AND EXTRACTION
  - CLASSIFICATION AND CLUSTERING
  - TRANSLATION
- MACHINE LEARNING
  - PREDICTIVE ANALYTICS
  - DEEP LEARNING

Weak/Narrow AI
Focuses on one specified task – no self-awareness or genuine intelligence. Siri, the voice-activated digital assistant developed by Apple, is a good example of a weak AI, combining several weak AI techniques (speech recognition, natural language processing, search algorithms, speech synthesis) to function. It can do a lot of things for the user, but fails when asked question outside the limits of application. Weak AI is already around us solving many specific problems. Google applications are driven by multiple algorithms and data mining and extraction.

Strong/General AI
Human-level AI, when a computer rivals a human brain, replicating all the main qualities and abilities. Strong AI would be able to perform all tasks that a human is capable of. Also includes the **hard problem of consciousness**: how and why we can think about thinking, self-identify and have integrated by distinct phenomenal experiences – how sensations acquire characteristics, such as the colour purple and the taste of grass.

THINKING ABOUT THINKING

One of the long-term goals of AI is to implement aspects of human intelligence on computers. Computational modeling uses simulations to study how human intelligence may be structured. There is some debate as to whether the mind is best concieved of as a huge array of small, individual components, such as neurons, or as an assembly of higher-level structures such as symbols, schemes, plans, and rules. The question remains: is it possible to accurately simulate a human brain on a computer without accurately simulating the neurons, the **connectome**, that make up the human brain.

KNOWLEDGE

Knowledge learned

Knowledge integrated

Knowledge retrieved and assembled

JUDGEMENT PROCESSOR

Judgements and intuitions

HUMANNESS/SAPIENCE

PERCEPTION AND CONCEPTION

CLEVERNESS

STRATEGIC PERSPECTIVE

DECISION MAKING

SYSTEMS PERSPECTIVE

MORAL SENTIMENTS

Modulation

Modulation

EMOTIONS

THE BINDING PROBLEM

Once the field of AI starts to consider emulating human experience using the information processing model of human thinking, it runs in to very complex problems of cognition (the mind and its processes) and sensation.

UNIFIED SUBJECTIVE EXPERIENCE

SELF

NOT-SELF

CONTROL

SPECULATION COMMON SENSE

ACTION

EMOTION

KNOWLEDGE

SENSORY

The **binding problem** is a term used at the interface between AI, neuroscience, cognitive science and philosophy of mind that has two main aspects.

Firstly, there is the **segregation problem**: a practical computational problem of how brains segregate elements in complex patterns of sensory input so that they are allocated to discrete objects. In other words, when looking at a book on a chair, what neural mechanisms ensure that the book and the chair are categorised as separate objects with distinct functions and properties? The segregation problem is sometimes called **BP1**.

Secondly, there is the **combination problem**: the problem of how segregated objects, backgrounds and abstract sensations and emotional features, are combined into a single unified subjective experience. The combination problem is sometimes called **BP2**.

Artificial superintelligence
A computer intellect smarter than the best human brains in practically all fields: scientific creativity, general wisdom and social intelligence.

H$_{2,500,000,000}$ O$_{970}$

N$_{47,000,000}$ P$_{9,000,000}$

Na$_{1,900,000}$ S$_{1,600,000}$

Fe$_{55,000}$ F$_{54,000}$ Zn$_{12,}$

Cr$_{98}$ Mn$_{93}$ Ni$_{87}$ Se$_{65}$

000,000 C490,000,000

Ca8,900,000 K2,000,000

Cl1,300,000 Mg300,000

00 Si9,100 Cu1,200 B710

Sn64 I60 Mo19 Co17 V1

The human molecule — the body reduced to a compound ... our body contains about 25 mg of vanadium

REFERENCES AND FURTHER READING

Pages 16–17 *Powers of ten* Inspired by the original and the best, Ray and Charles Eames' *Powers of Ten* (www. eamesoffice.com/the-work/powers-of-ten). See also *Cosmic View: The Universe in 40 Jumps*, by Kees Boeke (www. vendian.org/mncharity/cosmicview). Some images were adapted from www.atlasoftheuniverse.com; Google Earth provided the Earth images; NASA/ESA for Hubble imagery.

Pages 22–3 *The geometry of the Universe* Degrees of Freedom is a good blog with a couple of good posts on the geometry of space: blogs.scientificamerican.com/degrees-of-freedom/?page=3

Pages 24–5 *The unfurling fundamental forces* What the hell is the Higgs Boson? There is a good article here: good luck! (http://quantum-bits.org/?p=233)

Pages 28–9 *Matter and missing matter* The jury remains out on the greatest scientific mystery of our age (www. fnal.gov/pub/science/particle-physics/experiments/dark-matter-and-dark-energy.html)

Pages 30–1 *Big Bang to 380,000 years later* en.wikipedia. org/wiki/Recombination_(cosmology); Wilkinson Microwave Anisotropy Probe image of CMB: NASA/WMAP Science Team

Pages 34–5 *Relativity* Plots of LIGO gravitational waves modified from 'Observation of Gravitational Waves from a Binary Black Hole Merger', B. P. Abbott *et al.* (LIGO Scientific Collaboration and Virgo Collaboration) *Physical Review Letters* 116, 11 February 2016 (http://journals. aps.org/prl/abstract/10.1103/PhysRevLett.116.061102). See also (though now outdated by discoveries at LIGO) www.newscientist.com/article/dn25243-einsteins-ripples-your-guide-to-gravitational-waves.

Pages 36–7 *Atomic theory* en.wikipedia.org/wiki/Atomic_theory

Pages 38–9 *Quantum energy states* en.wikipedia.org/wiki/Quantum_state; en.wikipedia.org/wiki/Energy_level

Pages 40–1 *Star formation* Great gallery of images from the Chandra x-ray space observatory: chandra.harvard. edu/photo/category/stars.html

Pages 42–3 *Stellar nucleosynthesis* Some graphic elements modified from fission diagrams by commons. wikimedia.org/wiki/User:Borb

Pages 50–1 *Carbon* Allotrope models modified from work by Jozef Sivek: commons.wikimedia.org/wiki/File:Carbon_allotropes.svg

Pages 56–7 *Radioactive elements* Interactive chart of nuclides: www.nndc.bnl.gov/chart

Pages 58–9 *Nuclear fission* www.atomicarchive.com/Effects/effects1.shtml; en.wikipedia.org/wiki/TNT_equivalent; en.wikipedia.org/wiki/Energy_density

Pages 62–3 *Galaxies* For more information on the De Vaucouleurs system of classification, see en.wikipedia. org/wiki/Galaxy_morphological_classification. The Image of our galaxy modified from image by Andrew Z. Colvin.

Pages 64–5 *Black holes* www.nasa.gov/audience/forstudents/k-4/stories/nasa-knows/what-is-a-black-hole-k4.html

Pages 66–7 *The Sun* en.wikipedia.org/wiki/List_of_largest_stars. Graphic of the Sun modified from work by Kelvinsong on Wikimedia Commmons.

Pages 74–5 *Planet formation* http://spacemath.gsfc.nasa.gov/astrob/10Page7.pdf

Pages 78–9 *The Solar System: non-planetary bodies* Classification of Solar System bodies modified from work by Ariel Provost.

Pages 80–1 *The Moon and other moons* www.nasa.gov/pdf/58199main_Exploring.The.Moon.pdf

Pages 84–5 *The lithosphere*

Pages 86–7 *Plate tectonics* www.astrobio.net/news-exclusive/plate-tectonics-could-be-essential-for-life. Maps of ocean-floor spreading modified from work by Steven Dutch, Natural and Applied Sciences, University of Wisconsin/ www.uwgb.edu/dut

Pages 88–9 *Earthquakes* Seismic wave diagrams modified from illustrations in Introduction to Seismology (2nd edn), Peter M. Shearer, Cambridge University Press, 2009.

Pages 90–1 *Our atmosphere* Image of ozone hole NASA/NOAA

Pages 92–3 *Climate zones: atmospheric circulation* North Atlantic currents after *General Oceanography* (trans. H.U. Roll), Dietrich, G. et al., Willey, New York, 1980.

Pages 94–5 *Meteorology* Some of the diagrams inspired by the work of Lowell Hess – ground-breaking infographics from the 1950s.

Pages 96–7 *Climate extremes* Images of Mars and Venus – NASA.

Pages 98–9 *Solar activity and climate* solarscience.msfc.nasa.gov/SunspotCycle.shtml

Pages 100–1 *Climate change: snowball Earth* www.snowballearth.org/overview.html

Pages 106–7 *Proteins* Inestimable resource for information on proteins and protein structures: RCSB Protein Data Bank, or PDB (www.rcsb.org/pdb. Parts of protein structure diagram modified from work by Mariana Ruiz Villarreal: LadyofHats on Wikimedia.

Pages 116–17 *Abiogenesis* Ocean-floor diagram modified from work by Steven Dutch, Natural and Applied Sciences, University of Wisconsin/ www.uwgb.edu/dut. Vent diagram after Brazelton, W. J. et al., 'Methane- and sulfur-metabolizing microbial communities dominate the Lost City hydrothermal field ecosystem', Applied Environmental Microbiology, 72(9), 6257–70, 2006.

Pages 124–5 *Kingdoms of life* Phylogenetic tree of life modified from work by Ivica Letunic: Iletunic. Retraced by Mariana Ruiz Villarreal: LadyofHats.

Pages 132–3 *The Great Oxygenation Event* Debate: www.theguardian.com/science/2016/may/18/complex-life-on-earth-began-billion-years-earlier-than-previously-thought-study-argues. Banded iron formation data: Isley, A. E. and Abbott, D. H., 'Plume-related mafic volcanism and the deposition of banded iron formation', *Journal of Geophysical Research*, 104:15, 461–15, 1999.

Pages 134–5 *Photosynthesis and primary production* Circos diagram: http://mkweb.bcgsc.ca/tableviewer/

Pages 136–7 *Metabolic pathways* Pathways diagram adapted from KEGG Atlas: Kyoto Encyclopedia of Genes and Genomes –www.kegg.jp/kegg/atlas/?01100

Pages 138–9 *Enzymes* Diagram of conformation changes in hexokinase adapted from Bennet, W. S. and Steitz, T. A., *Journal of Molecular Biology*, 140, 211, 1980.

Pages 140–1 *The immune system: biological self-recognition* For more information there is a good Wikipedia page: en.wikipedia.org/wiki/Immune_system

Pages 142–3 *Chromosomes: structure and packing* Diagram of chromosome packing adapted from Pierce, Benjamin, *Genetics: A Conceptual Approach* (2nd edn) W.H. Freeman, New York, 2005.

Pages 144–5 *Cell reproduction* https://publications.nigms.nih.gov/insidethecell/chapter4.html

Pages 148–9 *Bacteria* Human microbiota diagram adapted from diagram at The National Human Genome Research Institute www.genome.gov/dmd/img.cfm?node=Photos/Graphics&id=85320

Pages 150–1 *Single-cell life* Good article about the transition from single-cell life to multicellular animals: www.wired.com/2014/08/where-animals-come-from. Background of comb jellies from *Kunstformen der Natur* (*Art Forms in Nature*) by Ernst Haeckel, 1904.

Pages 158–9 *The Cambrian explosion* Graphic of *Augaptilus filigerus* from Ernst Haeckel (commons.wikimedia.org/wiki/File:Haeckel_Copepoda.jpg). Some elements of graphic, modifications from Erwin, Douglas H. *et al.*, 'The Cambrian conundrum: Early divergence and later ecological success in the early history of animals', *Science*, 334:6059, 1091–7, 2011.

Pages 160–1 *The chordates* www.britannica.com/animal/chordate. en.wikipedia.org/wiki/Cephalization

Pages 162–3 *Fish* en.wikipedia.org/wiki/Jaw; www.palaeontologyonline.com

Pages 168–9 *Invasion of the land* A good article about terrestrialisation: www.paulselden.net/uploads/7/5/3/2/7532217/elsterrestrialization.pdf

Pages 170–1 *Carboniferous period* Research referenced: DiMichele, William A. and Philips, T. L., 'The response of hierarchically structured ecosystems to long term climate change: a case study using tropical peat swamps of Pennsylvanian age', in Stanley, Steven M., Knoll, A. H. and Kennett, J. P., *Effects of Past Global Change on Life* (Studies in Geophysics), National Research Council, 1995.

Pages 172–3 *Arthropods* Arthropod outlines modified from Jarmila Kukalová-Peck, *Palaeodiversity*, 2, 169–198, 2009. Chart of species modified Modified from Schminke, H. K. (after Wilson, 1992) 'Entomology for the copepodologist', *Journal of Plankton Research*, 29, i149-i162 (2007)

Pages 174–5 *Amniotes* finstofeet.com/2012/01/05/coming-of-the-amniotes/

Pages 178–9 *Trees and forests* Global forest cover data from UNEP GEO Data Portal, compiled from UNEP-WCMC.

Pages 180–1 *Pterosaurs: vertebrate flight* Interesting piece on pterosaur flight launching: lorenabarba.com/blog/student-guest-blog-post-pterosaur-quad-launch/. www.pteros.com

Pages 184–5 *Vertebrate vision en.wikipedia.org/wiki/Evolution_of_the_eye*

Pages 186–7 *The Chicxulub impactor* Peter Schulte, *et al.*, 'The Chicxulub Asteroid Impact and Mass Extinction at the Cretaceous-Paleogene Boundary', *Science*, 327, 1214, 2010.

Pages 188–9 *Bird migration* www.rspb.org.uk/discoverandenjoynature/discoverandlearn/funfactsandarticles/migration/

Pages 190–1 *The rise of the mammals* Felisa A. Smith, *et al.*, 'The Evolution of Maximum Body Size of Terrestrial Mammals', *Science*, 330, 1216, 2010.

Pages 192–3 *Gestation* en.wikipedia.org/wiki/Gestation_period

Pages 196–7 *Evolution of grasses* Data from Stromberg, Caroline A. E., 'Evolution of Grasses and Grassland Ecosystems', *Annual Review of Earth and Planetary Sciences*, 2011, 39:517–44.

Pages 198–9 *The biosphere and the carbon cycle* For an introduction: www.britannica.com/science/biosphere

Pages 222–3 *Human part-title* After Figure 2: http://rstb.royalsocietypublishing.org/content/royptb/369/1653/20130527.full.pdf

Pages 206–7 *Primates* There is a good Wiki on human evolution: en.wikipedia.org/wiki/Human_evolution

Pages 208–9 *Early humans* Laetoli footprints adapted from Agnew, Neville and Martha Demas, 'Preserving the Laetoli Footprints', *Scientific American*, September 1998, 47–9.

Pages 210–11 *The human diaspora* There is a good Wiki on early human migrations: en.wikipedia.org/wiki/Early_human_migrations

Pages 212–13 *Human anatomy* en.wikipedia.org/wiki/Anatomically_modern_human

Pages 220–1 *Language evolution* Data: http://visual.ly/writing-systems-world?view=true. Bouckaert, R. *et al.*, 'Mapping the origins and expansion of the Indo-European language family', *Science*, 337, 957–960, 2012. http://language.cs.auckland.ac.nz/media-material/

Pages 222–3 *Brain evolution: size and complexity* www.humanconnectomeproject.org

Pages 226–7 *Taste* The Coffee Taster's Flavor Wheel is licensed for use under the Creative Commons 4.0 International with book exemption, by the Specialty Coffee Association of America (SCAA/WCR).

Pages 228–9 *Neurotransmission* A good introduction to neurotransmission: www.mind.ilstu.edu/curriculum/neurons_intro/neurons_intro.php

Pages 232–3 *Learning and memory* www.education.com/reference/article/information-processing-theory/

Pages 234–5 *Computing* Visualisation using Raw at Density Design: http://raw.densitydesign.org

Pages 236–7 *Artificial Intelligence* Input from http://faculty.washington.edu/gmobus/TheoryOfSapience/SapienceExplained/3.sapiencecomponents/sapienceComponents.html. On the binding problem: en.wikipedia.org/wiki/Binding_problem The future? - en.wikipedia.org/wiki/Artificial_general_intelligence.

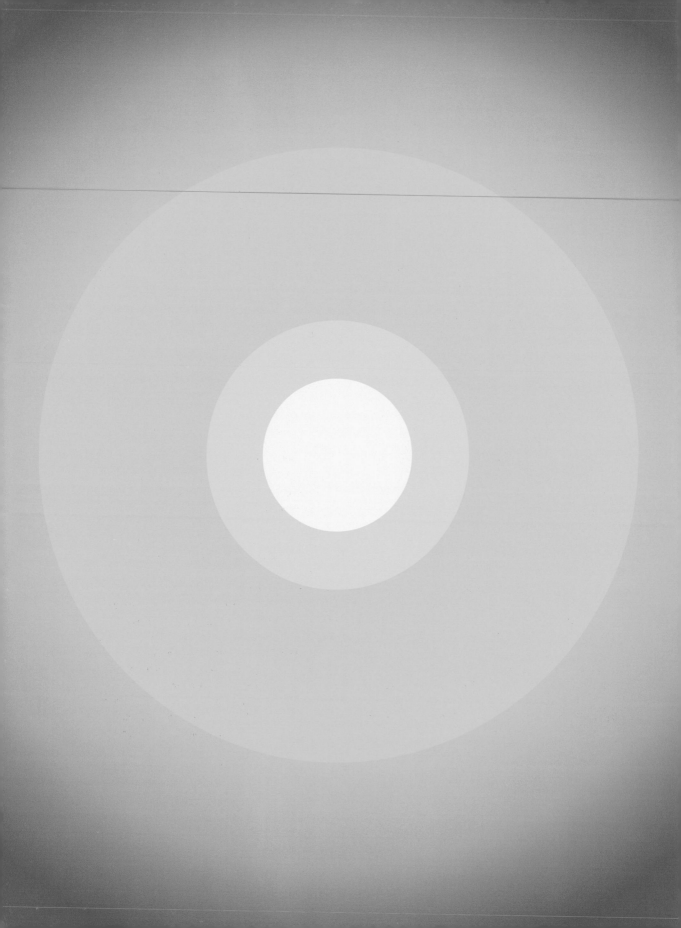

INDEX

O

Oberon (moon of Uranus) 77, 81

oceans: composition of 118–19; currents 93; marine habitats 164–5

oleic acid 105, 129

Oort Cloud 78, 79

Ordovician period 100, 156, 162, 168, 170, 172

organic chemistry 104–5

orogeny 86–7

osmium 52

osprey (*Pandion haliaetus*) 188

ostrich 174, 175, 176

ozone 91

P

Paedophryne amauensis 161

paleobiology 116–17, 118–19, 120–1, 124–5, 132–3, 146–7, 152–3, 156–7, 158–9, 160–1, 162–3, 168–9, 170–1, 172–3, 174–5, 176–7, 178–9, 180–1, 182–3, 184–5, 186–7, 188–9, 190–1, 194–5, 196–7, 206–7, 208–9, 210–11, 212–13 *see also under individual area of paleobiology*

Paleocene period 187, 190, 206

Paleocene-Eocene Thermal Maximum (PETM) 190

Paleogene Period 101, 157, 163, 171, 173, 183, 187, 190, 195, 206, 212

Panderichthys 169

paraceratherium 191

paramecium 150

particle-wave duality 36, 39

pedosphere 198

penguins 195

peptide 27, 106, 107, 116, 117, 122, 123, 229

perception 220, 224–5, 226–7, 236, 237

pederpes 169

Periodic Table 44–5

Permian Mass Extinction 157

Permian period 101, 157, 163, 170, 171, 173, 182, 194

pharyngeal gill slits 160

phase transitions 53

Phobos (moon of Mars) 76

phospholipid 105, 120, 121, 122, 129

phosphorylation 129

photons 22, 24, 25, 27, 28, 30, 31, 34, 38, 40, 43, 66, 67

photosynthesis 126, 132, 134–5, 150, 165, 198, 199

phylogeny 124

Pi 68–9

Pistol Star 67

placoderms 162

Planck scale 25

Planck time 24

planetary nebula 61, 67

planets/planetary science 74–5, 76–7, 80–5, 90–1, 98–9, 100–1, 102–3 *see also under individual area of planetary science*

platinum 49

Pleistocene period 191, 213

Plutchik's wheel of emotion 230

Pluto (dwarf planet) 77, 78, 79, 81

plutoids 79

plutonium 57, 58

pneumodesmus newmani 168

polarised light detection 225

polonium 210 57

polypeptide 106, 107

Precambrian Period 83, 132–3, 152–3

Primary waves (P-waves) 88

primates, evolutionary history of 206–7

proboscidea 191

prokaryotes 125, 126, 127, 149

promethium 56

proportions of recorded living species, relative 173

proteins 102, 104, 106–7, 108, 109, 114, 116, 121, 122, 123, 129, 130, 134, 138, 141, 143, 144, 150, 214, 215, 225, 227

Proteus (moon of Neptune) 81

protists 125, 150, 151

protocell 120–1

proton 14, 20, 24, 25, 26, 27, 28, 29, 30, 32, 37, 41, 43, 44, 56, 57, 58, 117, 120, 123, 128, 129

ACKNOWLEDGEMENTS

At William Collins, I would like to thank my publisher, Myles Archibald, and editor Julia Koppitz for sticking with me as the clock was ticking ... and ticking. Myles showed great faith in my ability to eventually deliver and great publishing skill in helping to tease the book out of me: a technique somewhere between breaking rocks and tickling a trout. Julia was a great support, particularly towards the latter stages of the project. It is a pleasure working with such an experienced and calm professional ... who is also a pleasure to know.

Thank you to my wife, Megan Smith, and daughters Edie and Rose, for putting up with a deranged hermit in the loft when they used to have a husband and father. I couldn't have made it without their love and forbearance. Megan also gave me vital visual input. And also to Kate McMillan, who gave me an early pep talk and sound advice: thanks.

My thanks too to the internet, without which a book like this simply wouldn't be possible. Thanks also go to some lifelong sources of visual joy and cross-disciplinary inspiration, Naum Gabo, Joan Miró, Gerhard Richter, Howard Hodgkin, Paul Klee, Eric Gill, and Ray and Charles Eames, who's prescient film *Powers of Ten* and subsequent book versions struck me profoundly, at an early age, with the power of the image to explain.

And lastly, to the great medieval Persian mathematician Muhammad ibn Musa al-Khwarizmi, for giving his name so stylishly to algorithms and, by association, logarithms, without the use of which my infographics simply wouldn't exist, and if they did, they wouldn't fit on the page.

Tom Cabot, June 2016

ABOUT THE AUTHOR

Tom Cabot is a London-based book editor and designer with a background in experimental psychology, natural science and graphic design. He founded the London-based packaging company, Ketchup, and has produced and illustrated many books for the British Film Institute, Penguin, Harper Collins and the Royal Institute of British Architects. Tom has long wanted to communicate the wonders of the Universe and the natural world graphically and accessibly – ever since being blown away by Ray and Charles Eames' *Powers of Ten* at an early age. This is his first book.